MAGNETOGASDYNAMICS AND PLASMA DYNAMICS

BY

SHIH-I PAI

RESEARCH PROFESSOR
INSTITUTE FOR FLUID DYNAMICS AND APPLIED MATHEMATICS,
UNIVERSITY OF MARYLAND, COLLEGE PARK, MARYLAND, U.S.A.

WITH 10 FIGURES

1962

SPRINGER-VERLAG / VIENNA

PRENTICE-HALL, INC. / ENGLEWOOD CLIFFS, N.J.

LIBRARY OF CONGRESS CATALOG CARD NUMBER 62-11113

PRINTED IN AUSTRIA

54552-C

TO

JOHANNES M. BURGERS

Preface

This book is based on the lecture notes which the author gave in a seminar of the same title in the Institut für theoretische Gasdynamik, D. V. L. e. V., Aachen, Germany, during the academic year of 1957-1958. The subject matter has been rewritten and expanded after the author's return to the University of Maryland. The purpose of this book is to give a theoretical introduction to plasma dynamics and magnetogasdynamics from the gasdynamic point of view. Attention is given to the basic assumptions and the formulation of the theory of the flow problems of a plasma, an ionized gas, as well as to the various methods of solving these problems. Since plasma dynamics is still in a developing stage, the author hopes that this book may furnish the readers some basic elements in the theory of plasma dynamics so that they may find it useful for further study and research in this new field.

After the introduction in which the scope of plasma dynamics is briefly discussed, the fundamental equations of plasma dynamics from the macroscopic point of view, i. e., the theory of continuum has been analyzed, in detail in chapters II to IV, including many simplified cases such as magnetogasdynamics, magnetohydrodynamics, electromagnetodynamics, radiation-magnetogasdynamics etc.

In chapter V, the important parameters and their range of applications have been treated. The parameters are useful in the correlation of experimental results.

In chapter VI to VIII, some simple flow problems of magnetohydrodynamics have been investigated while in chapters IX to XII, the flow problems of magnetogasdynamics are treated.

In chapters VI to XII, the plasma is considered as a single fluid of definite composition. In chapter XIII, deviations from such a theory of a single fluid have been discussed. These include ionization, plasma oscillation, particle motions etc.

In the last chapter a brief summary of plasma dynamics from the microscopic point of view, the kinetic theory of plasma, is given. The Boltzmann equation for plasma has been discussed. The relations between the kinetic theory and the continuum theory is also briefly stated.

The author wishes to express his sincere appreciation to Professor Johannes M. Burgers, to whom this book is dedicated and with whom the author has had the fortune to work for the last few years, for his inspiring discussions on various problems of plasma dynamics, to Professors K. Oswatitsch and M. H. Martin for their interest and encouragement and to Mr. A. I. Speth for reading the entire manuscript and checking the equations. In conclusion, the author takes this occasion to express his sincere gratitude to his daughter Sue Yu Pai who, since his first wife Chi Yu Feng Pai suddenly passed away in 1958, created a nice surrounding at the home, so that the author was able to finish the preliminary draft of this book and to his second wife Wang Yen-Lan for her constant encouragement and help in proof-reading during the revision and preparation of the final manuscript.

College Park, Maryland
October 1, 1961.

Shih-I Pai

Table of Contents

Chapter I

Introduction

1. Plasma and plasma dynamics. The study of flow problems of electrically conducting fluids, particularly of ionized gases, is currently receiving considerable interest. Such studies have been made for many years in connection with astrophysical (*2, 3*) and geophysical (*6*) problems such as sun spot theory, motion of the interstellar gas, origin of earth magnetism etc. Only recently some engineering problems need the studies of the flow of an electrically conducting fluid. For instance, the following are some of such problems:

(1) Controlled fusion research (*1, 9, 12*). Fusion reaction occurs only at very high temperature (the ignition temperature is of the order of $300 \cdot 10^6$ °K), at which the gas is in the plasma state. In fusion research, we use magnetic field to heat the cold plasma to the temperature above the ignition temperature and to confine the hot plasma. The stability of the flow of a plasma in a magnetic field is also very important for successful fusion reaction.

(2) Reentry problem of intercontinental ballistic missiles (*8*). The effects of electrical and magnetic fields on the heat transfer and drag of bodies in ionized gases during reentry are very important.

(3) Plasma jet (*7*). It is important in new propulsion unit for vehicles traveling in space. Water stabilized plasma jet is a good device to produce very high temperature gas flow and many interesting research may be performed.

(4) Communication. Radio wave in a plasma is important in the communication in space which differs considerably from the electromagnetic wave in a neutral gas or in a vacuum.

(5) Power conversion. The extraction of electrical energy directly from a hot plasma is a very interesting problem. It is so-called "magnetohydrodynamic generator" which has been extensively investigated by many authors (*15*).

Because of these engineering applications, many engineers and aerodynamicists joining the astrophysicists and geophysicists study extensively the dynamics of electrically conducting fluids.

At very high temperature, above 10^4 °K, the gas will be ionized. We may use the term "plasma" for ionized gas[1]. The properties of an ionized gas or a plasma differs considerably from "neutral" gas. Hence one may consider the plasma as the fourth state of matter. The most important difference of the

[1] Plasma was first used to represent a fully ionized gas in which the excess electric charge is zero. Since we are now interested in both the fully ionized gas and the partially ionized gas, the author prefers to generalize the definition of the plasma such that a plasma is an ionized gas.

properties of a plasma from those of an ordinary gas is that in a plasma the electromagnetic forces play a major role.

We should consider simultaneously electromagnetic forces and gasdynamic forces in our analysis of the flow problems of plasma. Many new phenomena occur due to the interaction of the gasdynamic and electromagnetic forces.

The science which deals with the flow phenomena of plasma is called "plasma dynamics." The scope of plasma dynamics is very broad. It contains problems from electrical discharge in rarefied gas, propagation of electromagnetic waves in ionized media to the so-called "magnetogasdynamics." Of course it includes ordinary gasdynamics and ordinary electrodynamics as special cases. It is impractical, if not impossible, to discuss all phases of plasma dynamics in a short book of this nature. We shall limit ourselves to problems where the electromagnetic forces are of the same order of magnitude as gasdynamic forces and their interaction is important, particularly for those problems known as "magnetogasdynamics." To be sure in the study of high temperature flow problem, the radiation field is not negligible. A complete analysis of plasma dynamics should consist of the study of the gasdynamic field, the electromagnetic field and the radiation field simultaneously. Little has been done on such a complete analysis. For simplicity, we shall assume that the radiation field may be determined from the parameter of gasdynamic and electromagnetic fields. Hence we shall discuss only the interaction of electromagnetic and gasdynamic fields with radiation field as a subsidiary factor. Furthermore we shall study this problem from the gasdynamic point of view. In other words, we would like to discuss how we could generalize our known results of various branches of fluid mechanics to include the effects of electromagnetic fields.

2. Methods of treatment. In gas dynamics, we are interested only in the macroscopic properties of the gas. There are two methods to discuss such macroscopic properties of gas: one is the statistical mechanics and the other is the theory of "continuum." In statistical mechanics we derive the properties of matter in bulk from the known atomic structure of matter and the laws of interaction between its constituents. For gas flow, we have to use the non-equilibrium statistical mechanics from which the kinetic theory of gases has been developed. In the theory of continuum, we postulate the macroscopic theory by considering the conservation laws, from which the ordinary theory of gas dynamics is developed. There are definite relations between these two approaches. It will be interesting to see how one could generalize the known results of these two methods for ordinary fluid mechanics to the case of plasma dynamics. This is one of the subjects of this book. For simplicity, we shall discuss our problem only from the classical mechanics point of view which is essentially the treatment of ordinary gas dynamics.

3. Microscopic point of view:—Kinetic theory of plasma. In the flow of ionized gas or plasma, there are positive and negative ions as well as electrons in the gas, which move between the neutral gas molecules and are continuously suffering collisions with them. The ions and electrons experience direct influences from the electric and the magnetic fields, which tend to produce an acceleration of their motion; owing to the collisions with the neutral molecules and with themselves, the ions may be retarded and in this way the forces acting on the ions and electrons become forces acting on the gas as a whole. An exact description of the phenomena occurring in the plasma requires a thorough investigation of collision phenomena. Such an analysis is known as kinetic theory of plasma. In ordinary kinetic theory of gases, we assume that the particles interact only during encounters, that they are substantially free between such encounters and that only binary collisions are considered. As a result, we may use one-particle distribution functions to describe the microscopic behavior of the systems. The resultant fundamental equation is the well-known Boltzmann equation. In a plasma, however, the particles are never free but interact permanently. A complete and satisfactory description of the reactions among the particles in a plasma is still not available. Usually we use essentially the method of ordinary kinetic theory of gases with one particle distribution function in the microscopic description of motion of a plasma. In this analysis, interaction terms, known as "Fokker-Planck" mechanism (*4*), are added to that of binary collision. The resultant equation may be called the generalized Boltzmann equation. This generalized Boltzmann equation should be solved simultaneously with Maxwell's equations for electric and magnetic fields. We shall very briefly discuss this microscopic description in chapter XIV.

The Boltzmann equation and its generalization are systems of non-linear partial-differentio-integral equations. At the present time it is not possible to solve the Boltzmann equation even for very simple flow problems. We would not expect to use the generalized Boltzmann equation to analyze flow problems for plasma in the near future. However, the Boltzmann equation serves two important aspects in the study of gasdynamics. In the first place, the fundamental equations of gasdynamics may be derived from Boltzmann equation as a first approximation. Thus we may have some guides about the validity of the fundamental equations for a macroscopic description from the analysis of the Boltzmann equation. In the second place, the Boltzmann equation may give us valuable information on the transport coefficients, such as the coefficients of viscosity, heat conductivity, etc. In the macroscopic analysis, these transport coefficients are simply introduced as known functions of physical quantities of gasdynamics. We would expect that the generalized Boltzmann equation would serve in a similar manner in the dynamics of plasma.

4. Macroscopic point of view:—Theory of Continuum. For many practical problems, the description of the motion of plasma in terms of Maxwell-Boltzmann distribution function is too detailed to be useful. We do not care about the motion of individual particles in the plasma and are interested only in the resultant effects due to the motion of a large number of particles. In other words, we are interested in the macroscopic quantities only, such as pressure, density, temperature, mean flow velocities, current density, etc. Here we postulate the fundamental equations for the dynamics of a plasma based on the conservation laws of mass, momentum, energy and charge. These fundamental equations should, of course, be consistent with those derived from the microscopic description. In the dynamics of a plasma, we should consider both the electromagnetic forces as well as other gasdynamic forces (*4, 10*).

In chapter II we shall derive the fundamental equations of plasma dynamics from the macroscopic point of view. In general the plasma should be considered as a mixture of n species: positive and negative ions, electrons and neutral particles. For each species we write down its "gasdynamic" equations with proper interaction terms due to electromagnetic forces. We should also use additional equations describing the electromagnetic phenomena with proper interaction terms due to gasdynamic motion.

In chapter III, we shall discuss the electromagnetic equations in some detail. The properties of the gasdynamic equations will be pointed out in the following chapters when we discuss the motion of the plasma.

5. Magnetogasdynamics and magnetohydrodynamics (*4, 5, 10, 13*). Even though plasma is a mixture of various species: charged and neutral particles, in many practical cases, the variation of its composition is small and its effect is negligible. In such cases, the plasma may be considered as a single fluid of definite composition. Then the fundamental equations of plasma may be greatly simplified. The resultant gasdynamic equations are very similar to the Navier-Stokes equations of gasdynamics but with interaction terms due to electromagnetic forces. Of course, these gasdynamic equations should be treated simultaneously with the equations of electromagnetic fields. Analysis based on such a system of equation may be called "Electro-Magneto-Gasdynamics," because we study the interaction of electromagnetic fields with gasdynamic forces. We shall derive this system of equations in chapter II.

There are many problems in which the energy in the electric field is much smaller than that in the magnetic field. In these cases, we may express all the electromagnetic quantities in terms of magnetic field. As a result, we shall consider only the interaction between the magnetic field and the gasdynamic field. This analysis is the well-known "Magnetogasdynamics." We shall discuss the magnetogasdynamic approximations and its resultat fundamental equations in chapter IV. The major portion of this book (chapter V to XII) treats various flow problems of magnetogasdynamics.

In the early days of investigations, we could not produce in the laboratory the flow phenomena of gases where the electromagnetic forces are of the same order of magnitude as the gasdynamic forces. Most of the investigations are theoretical. We could check these theoretical results with astronomical observations. However it was found that these interaction phenomena could be obtained in the laboratory by investigating the electromagnetic phenomena in the flow of a liquid such as mercury. Hence the term "Magnetohydrodynamics" has been found and extensively investigated. In magnetohydrodynamics, the compressibility effects of the medium is negligible. Magnetohydrodynamics has its own practical applications too. For instance, it may be used to deal with problems such as (1) cooling of nuclear reactor by liquid sodium and (2) induction flow meter which depends on the potential difference in the fluid in the direction perpenicular to the motion and to the magnetic field.

Now it is possible to produce the gas flow in which electromagnetic forces are of the same order of magnitude as those of gasdynamic forces by means of shock tube, electrical discharge or a combination of them. For problems mentioned in § 1, we should consider the effects of compressibility as well as the effects due to electro-magnetic forces. Hence we should study magnetogasdynamics instead of magnetohydrodynamics[1]. Just as in the case of ordinary gasdynamics, our results of magnetogasdynamics include the results of magnetohydrodynamics as a special case. We shall show how to derive the fundamental equations of magnetohydrodynamics from those of magnetogasdynamics. We shall discuss some problems of magnetohydrodynamics in chapters VI to VIII.

References

1. Bishop, A. S.: Project Sherwood. The U. S. Program in Controlled Fusion. Addison Wesley Publishing Co. Inc., 1958.
2. Burgers, J. M., and H. C. van de Hulst (ed.): Problems of Cosmic Aerodynamics. Central Air Documents Office, 1949.
3. Burgers, J. M., and H. C. van de Hulst (ed.): Gas Dynamics of Cosmic Clouds. Interscience Publishers, N.Y., 1955.
4. Clauser, F. H. (ed.): Plasma Dynamics. Addison Wesley Publishing Co. Inc., 1960.
5. Cowling, T. G.: Magnetohydrodynamics. Interscience Publishers, N.Y., 1957.
6. Elsasser, W. M.: Hydromagnetic Dynamo Theory. Rev. of Mod. Phys., vol. 28, pp. 135–163, April 1956.
7. Finkelnburg, W.: The Behavior of Matter at High Temperature. Symposium on High Temperature, Standard Research Institute, p. 39, 1956.
8. Kantrowitz, A. A.: A survey of Physical Phenomena Occurring in Flight at Extreme Speeds. Proc. Conf. on High Speed Aero. Poly. Inst. of Brooklyn, pp. 335–339, Jan. 1955.
9. Kurchatov, I. V.: Thermonuclear Fusion, Gas Discharge Research in the Soviet Union. Engineering, London, vol. 181, No. 4705, pp. 322–325, May 1956.
10. Landshoff, R. K. M. (ed.): Magnetohydrodynamics. A Symposium. Stanford University Press, 1957.
11. Pai, S. I.: Plasmadynamics from Gasdynamic Point of View. Proc. of First Intern. Symposium of Dynamics of Rarefied Gases, Nice, France, July 1958. Pergamon Press, pp. 394—405.
12. Post, R. F.: Controlled Fusion Research—An Application of the Physics of High Temperature Plasma. Rev. of Mod. Phys., vol. 28, No. 3, pp. 338–362, July 1956.
13. Resler, E. L. Jr., and W. R. Sears: The Prospects of Magneto-aerodynamics. Jour. Aero. Sci., vol. 25, No. 4, pp. 235–245, April 1958.
14. Spitzer, L. Jr.: Physics of Fully Ionized Gases. Interscience Publishers, N.Y., 1956.
15. Sutton, G. W.: Design Consideration of a Steady D. C. Magnetohydrodynamic Electric Power Generator. Tech. Inf. Ser. R 59-D-432, Aero-space Lab., Missile and Space Vehicle Dept., General Electric Co., 1959.

[1] The term "Magnetohydrodynamics" has been used in a very loose manner. The current practice seems that magnetohydrodynamics covers all the branches of plasma dynamics. It seems to the author that it is not right. The author prefers the terms "magnetogasdynamics" instead of "magnetohydrodynamics" in the study of the flow problems of an ionized gas (plasma).

Chapter II

Fundamental Equations of Plasma Dynamics

1. Introduction. The plasma may be considered as a mixture of n species which consists of ions, electrons and neutral particles. From macroscopic point of view, we would like to know the following quantities for each species:

T_s = (kinetic) temperature of sth species
p_s = pressure of the sth species
ρ_s = density of the sth species
u_s^i = ith component of the flow velocity of the sth species.

We would like also to know

E^i = ith component of the electric field, and
H^i = ith component of the magnetic field,

where $s = 1, 2, \ldots n$ and $i = 1, 2$ or 3. In plasma dynamics, there are $6n + 6$ unknown quantities to be investigated. It is assumed that all properties such as mass of a molecule of each species m_s and charge on a molecule of sth species e_s etc. are all known. In order to investigate these $6n + 6$ unknowns (T_s, p_s, ρ_s, u_s^i, E^i, H^i), we must find $6n + 6$ relations connecting these unknowns. These relations are $6n$ gasdynamic equations and 6 electromagnetic equations (*3, 6*). For each species, we have the following six gasdynamic equations:

(a) Equation of state which connects the temperature, the pressure and the density of each species. The existence of such a relation is an empirical fact. We shall discuss it in § 2.

(b) Equation of continuity which expresses the conservation of mass of each species. It will be discussed in § 3.

(c) Equations of motion which are generally three in number and express the relations of conservation of momentum in each species. We shall discuss these in § 4.

(d) Equation of energy which expresses the conservation of energy in each species which will be discussed in § 5.

Besides the above gasdynamic equations, we have the Maxwell's equations which are six in number and which describe the electromagnetic field. We shall give the Maxwell's equations in § 6. Further discussion of Maxwell's equations and other electromagnetic equations will be given in chapter III.

Even though it is sufficient to describe the flow of plasma by the "partial" quantities such as T_s, p_s, ρ_s, and u_s^i, it is sometimes more convenient to use some "gross" quantities which describe the physical properties of the plasma as a whole. We shall define these gross quantities in § 7. The fundamental equations which govern these gross quantities will be derived in §§ 8 to 10.

2. Equation of state. For each species of the plasma, there is a functional relation between p_s, ρ_s and T_s, i. e.,

$$p_s = f(\rho_s, T_s) \tag{2.1}$$

For plasma, the simple equation of state of a perfect gas is usually sufficient which may be written as

$$p_s = R_A \nu_s T_s \tag{2.2}$$

where R_A is the universal gas constant which is $1.381 \cdot 10^{-16}$ cm-dyne/° C. and ν_s is the number density of sth species such that $\rho_s = m_s \nu_s$ and m_s is the mass of a particle of sth species. Equation (2.2) is the first approximation of the more general virial equation of state (*4*).

3. Equation of continuity. The conservation of mass of sth species gives the equation of continuity of this species which may be written as

$$\frac{\partial \rho_s}{\partial t} + \frac{\partial}{\partial x^i}(\rho_s u_s^i) = \sigma_s \tag{2.3}$$

The summation convention is used for the repeated tensorial indices, i. e. i, but not for indices distinguishing the species "s". Here σ_s is the mass source per unit volume of sth species which may be due to ionization processes or other chemical reactions. If there is no chemical reaction and the degree of ionization of plasma is constant, σ_s is zero. Even if some of the σ_s are not equal to zero, by the conservation of mass of the plasma as a whole we have

$$\sum_{s=1}^{n} \sigma_s = 0 \tag{2.4}$$

Equation (2.4) is useful in deriving the equation of the plasma as a whole (§ 8).

4. Equation of motion. The equation of motion of sth species may be written as follows:

$$\frac{\partial \rho_s u_s^i}{\partial t} + \frac{\partial}{\partial x^j}(\rho_s u_s^i u_s^j - \tau_s^{ij}) = X_s^i + \sigma_s Z_s^i \tag{2.5}$$

where $\sigma_s Z_s^i$ is the ith component of the momentum source per unit volume associated with the mass source σ_s. We also have

$$\sum_{s=1}^{n} \sigma_s Z_s^i = 0 \tag{2.6}$$

The term τ_s^{ij} is the ijth component of the stress tensor of sth species. In the macroscopic analysis, we can only postulate the relation between τ_s^{ij} and other physical quantities. For instance, the simple relation is that τ_s^{ij} consists of pressure p_{ts} and viscous stress which may be assumed to be proportional to velocity gradient according to Navier-Stokes relation. Of course the stress τ_s^{ij} may also depend on temperature gradient or on concentration gradient. The partial pressure p_{ts} includes both the gasdynamic pressure p_s and the radiation pressure p_{Rs}. At low or at moderately high temperature, the radiation pressure is usually negligible. Only at very high temperatures the radiation pressure should be included in the analysis of flow problem.

The term X_s^i is the ith component of the body force per unit volume. It consists of the electromagnetic force and nonelectric body forces. We may write

$$X_s^i = F_{gs}^i + F_{es}^i \tag{2.7}$$

where F_{gs}^i is the ith component of the nonelectric forces such as gravitational force etc.

The electromagnetic force F_{es}^i may be written in the following form:

$$F_{es}^i = \rho_{es}\,[E^i + \mu_e\,(\vec{u}_s \times \vec{H})^i] + F_{es0}^i \tag{2.8}$$

where ρ_{es} is the electric charge density of sth species. $\rho_{es} = \nu_s\, e_s$ and e_s is the charge on a particle of sth species. μ_e is the magnetic permeability. $\vec{u}_s$ is the velocity vector of sth species with components u_s^i. The term F_{es0}^i is the interaction term, i. e., the force on the sth species due to all the other kind of particles in the plasma. By Newton's third law of motion, we have

$$\sum_{s=1}^{n} F_{es0}^i = 0 \tag{2.9}$$

The term $\vec{H}$ is the vector of magnetic field strength with component H^i.

5. Equation of energy. The conservation of energy of sth species gives the equation of energy of this species which may be written as

$$\frac{\partial \bar{e}_s}{\partial t} + \frac{\partial}{\partial x^j}(\bar{e}_s\, u_s^j - u_s^i\, \tau_s^{ij} - Q_s^j) = \varepsilon_s \tag{2.10}$$

where $\bar{e}_s$ = total energy of sth species in the plasma per unit volume $= \rho_s\, \overline{e_{ms}}$

$\overline{e_{ms}}$ = total energy of sth species of the plasma per unit mass

$$= U_{ms} + \frac{u_s^2}{2} + \Phi_s + \frac{E_{Rs}}{\rho_s} = \bar{I}_s + \frac{E_{Rs}}{\rho_s}$$

$$\bar{I}_s = U_{ms} + \frac{u_s^2}{2} + \Phi_s$$

U_{ms} = internal energy of sth species per unit mass

$\frac{u_s^2}{2}$ = kinetic energy per unit mass of sth species

Φ_s = potential energy per unit mass of sth species

E_{Rs} = radiation energy of sth species per unit volume.

The first term on the left hand side of equation (2.10) is the rate of change of energy of sth species per unit volume. The second term is the energy flow by convection, i. e., $\left(\frac{\partial}{\partial x^j}\bar{e}_s\, u_s^j\right)$. The third term $-\left(\frac{\partial}{\partial x^j} u_s^i\, \tau_s^{ij}\right)$ is the rate of energy dissipation by stress tensor τ_s^{ij}. The fourth term $-\left(\frac{\partial}{\partial x^j} Q_s^j\right)$ is the energy change due to hear flow, which consists of both heat conduction, Q_{cs}^j and radiation, Q_{Rs}^j; i. e., $Q_s^j = Q_{cs}^j + Q_{Rs}^j$.

The term on the right hand side of equation (2.10) is the energy production term or energy source per unit volume. It consists of terms due to electromagnetic field ε_{es} and due to chemical reaction ε_{cs}.

6. Maxwell's equations of electromagnetic field (7). The equations governing the electric and magnetic fields are the Maxwell's equations which are

$$\nabla \times \vec{H} = \vec{J} + \frac{\partial \vec{D}}{\partial t} \tag{2.11}$$

$$\nabla \times \vec{E} = -\frac{\partial \vec{B}}{\partial t} \tag{2.12}$$

where ∇ is the gradient operator. Gibbs vector notation is used. $\vec{H}$ is the vector magnetic field strength with components H^i, $\vec{E}$ is the vector electric field strength, $\vec{J}$ is the electric current density, t is the time and

$$\vec{D} = \varepsilon \vec{E} = \text{dielectric displacement} \tag{2.13}$$

where ε is the inductive capacity, sometimes referred to as the dielectric constant.

$$\vec{B} = \mu_e \vec{H} = \text{magnetic flux density} \tag{2.14}$$

where μ_e is the magnetic permeability.

Ordinarily we may assume that both ε and μ_e are constant for a given isotropic material. For anisotropic matter we should consider both ε and μ_e as tensor quantities and then $\vec{D}$ and $\vec{E}$ are not parallel nor $\vec{B}$ and $\vec{H}$. However in our analysis of plasma dynamics, we may assume that both ε and μ_e are constant. It is sufficient to take the values of ε and μ_e in vacuum (free space) as a first approximation. In this book, we always use these values in free space which in MKS or Giorgi system are respectively as follows:

$$\mu_e = 4\,\pi \cdot 10^{-7} \frac{\text{kg} \cdot \text{m}}{(\text{coulomb})^2} \tag{2.15}$$

and

$$\varepsilon = 8.854 \cdot 10^{-12} \frac{\text{coulomb}^2 \ \text{sec}^2}{\text{kg} \cdot \text{m}^3} \tag{2.16}$$

The interaction of the electromagnetic field equations with the flow of plasma is through the term $\vec{J}$ which is

$$J^i = \sum_{s=1}^{n} \rho_{es}\, u_s{}^i \tag{2.17}$$

The electric current density $\vec{J}$ is due to the motion of various charged particles in the plasma.

7. Gross variables of the plasma as a whole. Even though it is sufficient to use the "partial" quantities (ρ_s, p_s, T_s, $u_s{}^i$) to describe the motion of plasma, it is sometimes more convenient to use the "gross" quantities of the plasma as a whole in our analysis of the plasma dynamics. The relations between the gross quantities and the corresponding partial quantities are defined as follows:

(a) Density ρ of the plasma. The number density ν of the plasma is the sum of the number density ν_s of all the species, i. e.,

$$\nu = \sum_{s=1}^{n} \nu_s \tag{2.18}$$

The mass density or simply the density ρ of the plasma is

$$\rho = m\,\nu = \sum_{s=1}^{n} m_s\,\nu_s = \sum_{s=1}^{n} \rho_s \tag{2.19}$$

where m is the mean mass of a particle in the plasma.

(b) Pressure p of the plasma. The pressure p of the plasma is the sum of all the partial pressures of the species p_s in the plasma, i. e.,

$$p = \sum_{s=1}^{n} p_s \tag{2.20}$$

(c) Temperature T of the plasma. The "kinetic" temperature T of the plasma is defined as follows:

$$T = \frac{1}{\nu} \sum_{s=1}^{n} \nu_s\,T_s \tag{2.21}$$

(d) Flow velocity u^i of the plasma. The ith component of the flow velocity u^i of the plasma as a whole is defined by the following relation:

$$\rho\,u^i = \sum_{s=1}^{n} \rho_s\,u_s{}^i \tag{2.22}$$

The diffusion velocity $w_s{}^i$ of the sth species in the plasma is then

$$w_s{}^i = u_s{}^i - u^i \tag{2.23}$$

It is evident that

$$\sum_{s=1}^{n} \rho_s\,w_s{}^i = 0 \tag{2.24}$$

(e) Excess electric charge ρ_e. The sum of the charge densities ρ_{es} of all species gives the excess electric charge ρ_e of the plasma, i. e.,

$$\rho_e = \sum_{s=1}^{n} \rho_{es} = \sum_{s=1}^{n} e_s\,\nu_s \tag{2.25}$$

(f) Electrical current density J^i. The electrical current density in the plasma is defined as

$$J^i = \sum_{s=1}^{n} \rho_{es}\,u_s{}^i = \sum_{s=1}^{n} \rho_{es}\,w_s{}^i + u^i \sum_{s=1}^{n} \rho_{es} = i^i + \rho_e\,u^i \tag{2.26}$$

where i^i is the ith component of the conduction current and $\rho_e\,u^i$ is the ith component of the convection current.

8. Fundamental equations for the gross quantities of the plasma. The fundamental equations for the gross quantities of the plasma may be derived from these for the partial quantities. These equations for the gross quantities are given below:

(a) Equation of state of the plasma. If we add all the n equations of the type of equation (2.2), we have

$$p = \sum_{s=1}^{n} p_s = R_A \sum_{s=1}^{n} \nu_s T_s = R_A \nu T = R_p \rho T \tag{2.27}$$

where $R_p = R_A/m =$ the gas constant of the plasma. Equation (2.27) is the equation of state of the plasma as a whole.

(b) Equation of continuity of the plasma. If we add all the n equations of the type of equation (2.3) and use the relation (2.4), we have

$$\frac{\partial \rho}{\partial t} + \frac{\partial}{\partial x^i} (\rho u^i) = 0 \tag{2.28}$$

This is the equation of continuity of the plasma.

(c) The equations of motion of the plasma. If we add all the n equations of the type of equation (2.5), we have

$$\frac{\partial \rho u^i}{\partial t} + \frac{\partial \rho u^i u^j}{\partial x^j} = \rho \left(\frac{\partial u^i}{\partial t} + u^j \frac{\partial u^i}{\partial x^j}\right) = \rho \frac{D u^i}{D t} = -\frac{\partial p_t}{\partial x^i} + \frac{\partial \tau^{ij}}{\partial x^j} + F_e^i + F_g^i \tag{2.29}$$

This is the equation of motion of the plasma in the direction of the ith axis. The various terms in equation (2.29) are explained as follows:

The term F_g^i is the nonelectric force, i. e.,

$$F_g^i = \sum_{s=1}^{n} F_{gs}^i \tag{2.30}$$

The term F_e^i is the electromagnetic force, i. e.,

$$F_e^i = \sum_{s=1}^{n} F_{es}^i = \rho_e E^i + \mu_e (\vec{J} \times \vec{H})^i \tag{2.31}$$

The stress tensor τ_s^{ij} of the sth species may be written as

$$\tau_s^{ij} = \tau_{s0}^{ij} - p_{ts} \delta^{ij} \tag{2.32}$$

where $\delta^{ij} = 0$ if $i \neq j$ and $\delta^{ij} = 1$ if $i = j$. τ_{s0}^{ij} is the ijth component of the viscous stress and p_{ts} is the total pressure which includes both the gasdynamic pressure p_s and the radiation pressure p_{Rs}.

The total stress tensor τ^{ij} of the plasma is

$$\tau^{ij} = \sum_{s=1}^{n} \tau_{s0}^{ij} - \sum_{s=1}^{n} \rho_s w_s^i w_s^j \tag{2.33}$$

The stress tensor is also due to a very complicated molecular motion. For first approximation, we may assume that τ^{ij} is of the following form

$$\tau^{ij} = \mu \left(\frac{\partial u^i}{\partial x^j} + \frac{\partial u^j}{\partial x^i}\right) + \mu_1 \frac{\partial u^k}{\partial x^k} \delta^{ij} \tag{2.34}$$

where μ is the ordinary coefficient of viscosity and μ_1 is the second coefficient of viscosity. Ordinarily we may also assume that $\mu_1 + \frac{2}{3}\mu = 0$. The coefficients of viscosity are functions of temperature and the composition of the plasma. They are also functions of the electric and magnetic fields (*1, 4*).

The total pressure consists of the gasdynamic pressure p and the radiation pressure p_R, i. e.,

$$p_t = \sum_{s=1}^{n} p_{ts} = \sum_{s=1}^{n} p_s + \sum_{s=1}^{n} p_{Rs} = p + p_R \tag{2.35}$$

The gasdynamic pressure is the value of pressure used in the equation of state. The radiation pressure is determined by a very complicated radiation phenomenon. However for engineering problem, we may use the following simple formula (*2*)

$$p_R = \frac{a_R}{3} T^4 \tag{2.36}$$

where a_R is known as Stefan-Boltzmann constant $= 7.67 \cdot 10^{-15}$ ergs-cm^{-3} deg. C^{-4}.

(d) Equation of energy. If we add all the n equations of the type of (2.10) we have

$$\frac{\partial \rho \overline{e_m}}{\partial t} + \frac{\partial \rho \overline{e_m} u^j}{\partial x^j} = -\frac{\partial u^j p_t}{\partial x^j} + \frac{\partial u^i \tau^{ij}}{\partial x^j} + \frac{\partial Q^j}{\partial x^j} + \varepsilon_t \tag{2.37}$$

This is the energy equation of the plasma. The various terms in equation (2.37) are explained as follows:

The total energy of the plasma per unit mass is

$$\overline{e_m} = U_m + \frac{u^2}{2} + \Phi + \frac{E_R}{\rho} = \bar{I} + \frac{E_R}{\rho} \tag{2.38}$$

where $\bar{I} = U_m + \dfrac{u^2}{2} + \Phi$.

$U_m = \dfrac{1}{\rho} \sum_{s=1}^{n} (\rho_s U_{ms} + \frac{1}{2} \rho_s w_s^2) =$ total internal energy of the plasma per unit mass.

$\Phi = \dfrac{1}{\rho} \sum_{s=1}^{n} \Phi_s \rho_s =$ total potential energy of the plasma per unit mass.

u is the resultant velocity of the plasma as a whole with components u^i.

$E_R = \sum_{s=1}^{n} E_{Rs} =$ total radiation energy of the plasma per unit volume. As a first approximation which is consistent with equation (2.36), we have

$$E_R = a_R T^4 \tag{2.39}$$

The heat flow Q^j of the plasma consists of both heat conduction Q_c^j and radiation Q_R^j, i. e.,

$$Q^j = Q_c^j + Q_R^j \tag{2.40}$$

where the heat conduction flux is

$$Q_c^j = \sum_{s=1}^{n} (Q_{cs}^j + u^i \rho_s w_s^i w_s^j - \rho_s \bar{I}_s w_s^j + \tau_{s0}^{ij} w_s^i - p_s w_s^j) \tag{2.41}$$

The heat conduction flux is due to a very complicated molecular motion. For first approximation, we may assume that

$$Q_c^j = \varkappa \frac{\partial T}{\partial x^j} \tag{2.42}$$

where $\varkappa$ is the coefficient of heat conductivity which is a function of temperature and composition of the plasma. It may also depend on the electric and magnetic field (*1*, *4*).

The radiation flux is

$$Q_R^j = \sum_{s=1}^{n} (Q_{Rs}^j + E_{Rs} w_s^j - p_{Rs} w_s^j) \tag{2.43}$$

The radiation flux is due to complex radiation phenomena. For a first approximation, we may assume that (*5*)

$$Q_R^j = D_R \frac{\partial E_R}{\partial x^j} \tag{2.44}$$

where D_R is known as the diffusion coefficient of radiation and may be written as

$$D_R = \frac{c l_R}{3} \tag{2.45}$$

$$l_R = \frac{1}{K_R \rho} = \text{Roseland mean free path of radiation} \tag{2.46}$$

and K_R is the opacity or Roseland mean absorption coefficient. K_R depends on both temperature T and density ρ. Here c is the velocity of light.

The energy source ε_t consists of one due to chemical reaction ε_c and one due to electromagnetic fields ε_e, i. e.,

$$\varepsilon_t = \sum_{s=1}^{n} \varepsilon_s = \sum_{s=1}^{n} (\varepsilon_{se} + \varepsilon_{sc}) = \varepsilon_e + \varepsilon_c \tag{2.47}$$

and

$$\varepsilon_e = E^j \cdot J^j \tag{2.48}$$

(e) Equation of conservation of electrical charge. If we multiply the factor $\gamma_s = e_s/m_s$ to equation (2.3), we have the equation of conservation of charge of each charged species which gives

$$\frac{\partial \rho_{es}}{\partial t} + \frac{\partial}{\partial x^j} (\rho_{es} u_s^j) = \gamma_s \sigma_s \tag{2.49}$$

The conservation of total electrical charge of the plasma gives

$$\sum_{s=1}^{n} \gamma_s \sigma_s = 0 \tag{2.50}$$

If we add all the n equations of the type of equation (2.49), we have the equation of conservation of electrical charge of the plasma or the electrical equation of continuity of the plasma. which is

$$\frac{\partial \rho_e}{\partial t} + \frac{\partial J^i}{\partial x^i} = 0 \tag{2.51}$$

(f) Equation of electrical current. Equation (2.5) multipled by γ_s gives

$$\frac{\partial \rho_{es} u_s^i}{\partial t} + \frac{\partial}{\partial x^j} (\rho_{es} u_s^i u_s^j - \gamma_s \tau_s^{ij}) = \gamma_s (X_s^i + \sigma_s Z_s^i) \tag{2.52}$$

If we add all the n equations of the type of equation (2.52), we have

$$\frac{\partial J^i}{\partial t} + \frac{\partial}{\partial x^j} \sum_{s=1}^{n} (\rho_{es} u_s^i u_s^j - \gamma_s \tau_s^{ij}) = \sum_{s=1}^{n} \gamma_s (X_s^i + \sigma_s Z_s^i) \tag{2.53}$$

Equation (2.53) is the differential equation for the current density J^i. In general equation (2.53) is too complicated to be analyzed. Usually in the analysis of plasma dynamics, a simple relation is used instead of equation (2.53). We often use the so-called generalized Ohm's law to replace equation (2.53). The generalized Ohm's law is

$$i^i = \sigma [E^i + \mu_e (\vec{u} \times \vec{H})^i] \tag{2.54}$$

where σ is known as the electrical conductivity. We shall discuss the relation between equation (2.53) and (2.54) in more detail in chapter III.

9. Fundamental equations of plasma dynamics. There are two ways to describe the motion of plasma:

(a) We may use the partial variables (T_s, p_s, ρ_s, u_s^i, E^i, H^i) to describe the motion and the states of the plasma where $s = 1, 2, \ldots n$ and $i = 1, 2, 3$. The fundamental equations in this case are equations (2.2), (2.3), (2.5), (2.10), (2.11), and (2.12). There are $6n + 6$ variables and $6n + 6$ equations. For instance there are n equations of the type of (2.2) and $3n$ equations of the type of equation (2.5).

(b) We may use the gross variables (T, p, ρ, u^i, E^i, H^i) and partial variables (T_r, p_r, ρ_r, u_r^i) where $r = 1, 2, \ldots n - 1$. The fundamental equations are (2.11), (2.12), (2.27), (2.28), (2.29), (2.37), and the corresponding equations for the partial variables. We may also replace the variables u_r^i by w_r^i and replace (2.5) by a corresponding equation of w_r^i. This approach is of great interest because the fundamental equations of magnetogasdynamics are derived from this approach.

Let us consider the case of fully ionized plasma in which $n = 2$. We have then 18 variables, i. e., T, p, ρ, u^i, E^i, H^i, T_1, p_1, ρ_1, w_1^i, where subscript 1 represents the values of one of the species in the plasma, e. g., those of the electrons. It is possible to replace the variables ρ_1 and w_1^i by ρ_e and J^i respectively (see chapter III). The fundamental equations for the flow of fully ionized plasma are equations (2.11), (2.12), (2.27), (2.28), (2.29), (2.37), and equations (2.2), (2.10), (2.51), and (2.53). For the last four equations we put s equal to 1 only.

10. Fundamental equations of electro-magneto-gasdynamics. As we mention in chapter I, § 5, in many practical problems we may consider the plasma as a single fluid. In this case, our variables are quite similar to those of fully ionized plasma, except that we are not interested in the partial values T_1 and p_1. There are now only 16 variables in our problem, i. e., T, p, ρ, u^i, E^i, H^i, ρ_e and J^i. The fundamental equations are then:

$$\nabla \times \vec{H} = \vec{J} + \frac{\partial \varepsilon \vec{E}}{\partial t} \tag{2.11}$$

$$\nabla + \vec{E} = -\frac{\partial \mu_e \vec{H}}{\partial t} \tag{2.12}$$

$$p = R_p \rho T \tag{2.27}$$

$$\frac{\partial \rho}{\partial t} + \frac{\partial}{\partial x^i} (\rho u^i) = 0 \tag{2.28}$$

$$\rho \frac{D u^i}{D t} = -\frac{\partial p_t}{\partial x^i} + \frac{\partial \tau^{ij}}{\partial x^j} + F_e{}^i + F_g{}^i \tag{2.29}$$

$$\frac{\partial \rho \overline{e_m}}{\partial t} + \frac{\partial \rho \overline{e_m u^j}}{\partial x^j} = -\frac{\partial u^j p_t}{\partial x^j} + \frac{\partial u^j \tau^{ij}}{\partial x^j} + E^j J^j + \frac{\partial Q^j}{\partial x^j} \tag{2.37}$$

$$\frac{\partial \rho_e}{\partial t} + \frac{\partial J^j}{\partial x^j} = 0 \tag{2.51}$$

$$J^i = i^i + \rho_e u^i = \sigma [E^i + \mu_e (\vec{u} \times \vec{H})^i] + \rho_e u^j \tag{2.55}$$

In this systems of equations we neglect the energy source due to chemical reaction and use the simple current equation (2.54) to replace the more accurate and also more complicated electrical current equation (2.53). Further approximations should be made in order to get the fundamental equations of magnetogasdynamics. We shall discuss these approximations in chapter IV.

References

1. Burgers, J. M.: Applications of Transfer Equations to the Calculation of Diffusion, Heat Conduction, Viscosity and Electric Conductivity. I and II. Technical Notes 124a and b. Institute for Fluid Dynamics and Applied Math. Univ. of Maryland, May 1958.
2. Chandrasekhar, S.: An Introduction to tho Study of Stellar Structure. Dover Publications, Inc., 1957.
3. Grad, H.: General Fluid Equations. Notes on Magnetohydrodynamics. No. 1. Inst. of Math. Sciences, New York Univ., 1956.
4. Hirschfelder, J. O., C. F. Curtiss, and R. B. Bird: Molecular Theory of Gases and Liquids. John Wiley & Sons, Inc., N.Y., 1954.
5. Marshak, R. E.: Effect of Radiation on Shock Wave Behavior. The Physics of Fluids, vol. 1, No. 1, Jan.–Feb., 1958, pp. 24–29.
6. Pai, S. I.: The Fundamental Equations of Magnetogasdynamics. Engineering report No. 10053-M, The Martin Company, Baltimore, April 1958.
7. Stratton, J. A.: Electromagnetic Theory. McGraw-Hill Book Co., Inc., N.Y., 1941.

Chapter III

Electromagnetic Equations in Plasma Dynamics

1. Introduction. The fundamental equations of plasma dynamics discussed in chapter II, § 9 may be roughly divided into two groups: one deals with mainly the gasdynamic quantities (T, p, ρ, u^i or their partial values) which may be called the gasdynamic equations; and the other deals with mainly the electromagnetic quantities (E^i, H^i, ρ_e, J^i) which may be called the electromagnetic equations. Of course, it should be noted that there are interaction terms in these two groups which are very important in our treatment and that the electromagnetic equations of ρ_e and J^i are derived from the gasdynamic equations as shown in chapter II, § 8. However it is sometimes convenient to make such a division because in this way we may easily compare our results of plasma dynamics with those of ordinary gasdynamics and electrodynamics. The properties of the gasdynamic equations will be discussed in the following chapters because one of our main purposes of this book is to treat the flow problems of plasma which are essentially gasdynamical in nature. In this chapter we shall briefly discuss the properties of electromagnetic equations and the approximations which will be used in our analysis.

In § 2, Maxwell's equations of electromagnetic field will be discussed while in § 3, the conservation of electrical charge will be discussed.

The electric current is due to a very complicated molecular motion. Usually we replace the more accurate equation (2.53) for J^i by the simple generalized Ohm's law (2.55) as a first approximation. In § 4, we shall show how we can derive equation (2.55) from (2.53) or its equivalent equations for $w_s{}^i$. In equation (2.55), we assume that the conduction current density i^i depends essentially on the electric and the magnetic fields. But from equation (2.53) it is evident that i^i must depend also on other gasdynamic quantities such as the pressure gradient etc. We shall briefly discuss these influences in § 4, too.

In the analysis of Maxwell's equations of electromagnetic fields, it is sometimes convenient to introduce vector and scalar potentials which will be discussed in § 5.

For every particular problem of plasma dynamics, we have to give certain initial and boundary conditions. Our problem is to find solutions of the fundamental equations which also satisfy these initial and boundary conditions (*6*).

By initial conditions we mean the values of all the variables (T, p, etc.) of the plasma dynamics at certain initial time $t = 0$. Customarily in plasma dynamics we do not give the spatial distributions of these initial values but we only require that the initial conditions be consistent with the boundary conditions for $t = 0$ and the fundamental equations.

By boundary conditions, we mean the values of these variables of plasma dynamics on the boundary of the domain investigated at all the time $t \geqq 0$. In plasma dynamics, we have the boundary conditions for both the gasdynamic and the electromagnetic variables. For gasdynamic variables, the usual non-slip condition will be used (*6*). In § 6, we shall discuss the boundary conditions for the electromagnetic fields.

There are a number of different systems of units in electromagnetic theory in common use. As a result, the same equations, e. g., Maxwell's equations, have different forms in different books or papers. In this book, the MKS or Giorgi system will be used. However we shall briefly summarize in § 7 those common systems of units used in the electromagnetic theory so that the reader may transform our equations into the form with which he is familiar.

2. Maxwell's equations. The Maxwell's equations for electromagnetic fields used in plasma dynamics (cf. chapter II, § 6) are

$$\nabla \times \vec{H} = \vec{J} + \frac{\partial \varepsilon \vec{E}}{\partial t} \tag{3.1}$$

$$\nabla \times \vec{E} = -\frac{\partial \mu_e \vec{H}}{\partial t} \tag{3.2}$$

where we assume μ_e and ε to be constant and of their values in free space given by equations (2.15) and (2.16) respectively. Here MKS or Giorgi system is used (cf. § 7).

In plasma dynamics, the following three conditions are usually satisfied:

(a) The flow velocity $\vec{u}$ is much smaller than the light velocity c, i. e.,

$$R_c = \frac{u^2}{c^2} \ll 1 \tag{3.3}$$

where the light velocity c is equal to

$$c = \frac{1}{\sqrt{\varepsilon \mu_e}} \cong 3 \cdot 10^{10} \text{ cm./sec. in free space} \tag{3.4}$$

(b) The electric field $\vec{E}$ is of the same order of magnitude as the induced electric field $\mu_e \vec{u} \times \vec{H}$, i. e.,

$$\vec{E} = 0\,(\mu_e \vec{u} \times \vec{H}) \tag{3.5}$$

(c) The problems of very high frequency will not be considered. The time scale t_0 is of the order of L/U where L is the typical dimension of the flow field and U is the typical velocity of the flow, i. e.,

$$R_t = \frac{t_0\, U}{L} = 0\,(1) \tag{3.6}$$

Under these three conditions, it can be shown that the displacement current $\partial \varepsilon \vec{E}/\partial t$ may be neglected in the Maxwell's equations (see chapter IV). The equations for electromagnetic fields become

$$\nabla \times \vec{H} = \vec{J} \tag{3.7a}$$

$$\nabla \times \vec{E} = -\frac{\partial \mu_e \vec{H}}{\partial t} \tag{3.7b}$$

Equation (3.7) may be called pre-Maxwell equations (*1*) for electromagnetic fields. We shall use equation (3.7) in most of our problems in this book because three conditions listed above are usually satisfied.

3. Equation of conservation of electrical charge (*5*). The equation of conservation of electrical charge of the plasma is given by equation (2.51) which may be written in the following form by the help of relation (2.55):

$$\frac{\partial \rho_e}{\partial t} + \frac{\partial}{\partial x^j}(\rho_e u^j) = -\frac{\partial i^j}{\partial x^j} \tag{3.8}$$

It is interesting to notice the similarity of equation (3.8) with the gas dynamic equation of continuity (2.28). The only difference occurs on the right hand side because in equation (2.28) we assume that there is no source nor sink of mass for the plasma as a whole while in equation (3.8), the divergence of conduction current $\vec{i}$ represents a source or sink of the electric charge in the plasma.

4. Equation of electrical current. We would like to find some simple relation of conduction current $\vec{i}$, i. e.,

$$i^i = \sum_{s=1}^{n} \rho_{es} w_s{}^i \tag{3.9}$$

in terms of some other quantities of plasma. In order to bring out the essential feature of this complicated molecular motion, we shall consider first a simple case of fully ionized plasma which consists of electrons and positive ions only. The mass of the ions m_2 is much larger than the mass of the electrons m_1, where subscript 1 refers to the values of electrons while subscript 2 refers to values of ions. We further assume that the plasma is approximately neutral, i. e.,

$$\rho_e = \rho_{e1} + \rho_{e2} = e_1 \nu_1 + e_2 \nu_2 \cong 0 \tag{3.10}$$

We consider that the ions are singly charged, $e_2 = -e_1$. Then

$$\nu_1 \cong \nu_2 \tag{3.11}$$

The number density of the electrons is approximately equal to that of ions. In this case, the mass density of ions is much larger than that of electrons,

$$\rho_2 = m_2 \nu_2 \gg m_1 \nu_1 = \rho_1 \tag{3.12}$$

The density of the plasma is approximately equal to that of the ions,

$$\rho = \rho_1 + \rho_2 \cong \rho_2 \tag{3.13}$$

The relation the diffusion velocities (2.24) gives

$$\rho_1 w_1{}^i + \rho_2 w_2{}^i = 0 \tag{3.14}$$

Then

$$|\, w_1{}^i \,| \gg |\, w_2{}^i \,| \tag{3.15}$$

The conduction current i^i is

$$i^i = \rho_{e1}\, w_1{}^i + \rho_{e2}\, w_2{}^i \cong \rho_{e1}\, w_1{}^i \tag{3.16}$$

Equation (3.16) shows that in the present case the conduction current is determined essentially by the diffusion velocity of the electrons.

From equations (2.5) and (2.29) with the help of equation (2.3) and (2.28), we have the equation for the diffusion velocity $w_r{}^i = u_r{}^i - u^i$ as follows:

$$\begin{aligned} &\frac{\partial w_r{}^i}{\partial t} + u_r{}^j \frac{\partial u_r{}^i}{\partial x^j} - u^j \frac{\partial u^i}{\partial x^j} = \\ &= \frac{1}{\rho_r}\cdot\frac{\partial \tau_r{}^{ij}}{\partial x^j} + \frac{1}{\rho}\cdot\frac{\partial p_t}{\partial x^i} - \frac{1}{\rho}\cdot\frac{\partial \tau^{ij}}{\partial x^j} + \frac{X_r{}^i}{\rho_r} - \frac{F_e{}^i + F_g{}^i}{\rho} + \frac{\sigma_r}{\rho_r}(Z_r{}^i - u_r{}^i) \end{aligned} \tag{3.17}$$

If we want to find a relation for conduction current $\vec{i}$ under the conditions that it is explicitly independent of the time t and spatial coordinates x^j, with the electromagnetic forces as the only predominant force and with no source term, i. e., $\sigma_r = 0$, equation (3.17) reduces to the following simple form (*4*):

$$\frac{F_{er}{}^i}{\rho_r} = \frac{F_e{}^i}{\rho} \tag{3.18}$$

or

$$\begin{aligned} &(\rho\, \rho_{er} - \rho_r\, \rho_e)\, \vec{E}_u + \mu_e\, (\rho\, \rho_{er}\, \vec{w}_r - \rho_r\, \vec{i}) \times \vec{H} = \\ &= -\rho\, \vec{F}_{er0} = -\rho \sum_{s=1}^{n} \alpha_{rs}\, (\vec{u}_r - \vec{u}_s) = -\rho \sum_{s=1}^{n} \alpha_{rs}\, (\vec{w}_r - \vec{w}_s) \end{aligned} \tag{3.19}$$

where the interaction forces between different charged particles are assumed to be proportional to the difference of their velocities $\vec{u}_r$ and $\vec{u}_s$ with corresponding proportional constant α_{rs}.

$$\vec{E}_u = \vec{E} + \mu_e\, \vec{u} \times \vec{H} \tag{3.20}$$

$\vec{E}_u$ is the electric field measured in a system moving with the plasma. If we consider ρ_r, ρ_{er}, $\vec{E}_u$, $\vec{H}$, and α_{rs} as given, equation (3.19) gives a set of linear equations for $\vec{w}_r$ because $\vec{i}$ is a linear function of $\vec{w}_r$. We may solve for $\vec{w}_r$ from equation (3.19). For our simple case of fully ionized plasma, equation (3.19) after neglecting small terms gives

$$\vec{i} = \sigma\, \vec{E}_u - \sigma\, \zeta\, \mu_e\, \vec{i} \times \vec{H} \tag{3.21}$$

where $\sigma = -\rho_{e1}{}^2/\alpha_{12}$ = electrical conductivity, and $\zeta = -1/\rho_{e1}$. The term $\zeta\, \mu_e\, \vec{H}$ is known as the Hall vector and $(\sigma\, \zeta\, \mu_e\, \vec{i} \times \vec{H})$ is the Hall current. When the Hall current is negligible, equation (3.21) reduces to the generalized Ohm's law:

$$\vec{i} = \sigma\, \vec{E}_u = \sigma\, (\vec{E} + \mu_e\, \vec{u} \times \vec{H}) \tag{3.22}$$

In many flow problems of plasma dynamics, the Hall current is negligibly small (cf. chapter XIII, § 6). Equation (3.22) has been generally used in plasma dynamics, particularly in magnetogasdynamics. From our analysis we see that the generalized Ohm's law (3.22) is a very much simplified approximation for the actual complicated current equation.

Equation (3.21) may be written in the form

$$\vec{i} = \sigma \vec{E}^{\parallel} + \frac{R}{R^2 + \zeta^2 B^2} \vec{E}_u{}^{\perp} - \frac{\zeta}{R^2 + \zeta^2 B^2} \vec{E}_u \times \vec{B} \tag{3.23}$$

where $R = 1/\sigma$, $\vec{E}_u = \vec{E}_u{}^{\parallel} + \vec{E}_u{}^{\perp}$, $\vec{E}_u{}^{\parallel} = \vec{B}(\vec{B} \cdot \vec{E}_u)/B^2$, $\vec{E}_u{}^{\perp} = \vec{B} \times (\vec{E}_u \times \vec{B})/B^2$, $\vec{B} = \mu_e \vec{H}$. The notations $\vec{E}_u{}^{\parallel}$ and $\vec{E}_u{}^{\perp}$ were first introduced by Grad (*4*). If $\vec{B}$ is parallel to $\vec{E}_u$, we have $\vec{E}_u = \vec{E}_u{}^{\parallel}$, and if $\vec{B}$ is perpendicular to $\vec{E}_u$ we have $\vec{E}_u = \vec{E}_u{}^{\perp}$. We may interpret that σ is the electrical conductivity in the direction of $\vec{B}$; $R/(R^2 + \zeta^2 B^2)$, the electrical conductivity perpendicular to $\vec{B}$; and the last term, the Hall current. In the present case, we see that for large $|\vec{B}|$, the conductivity perpendicular to $\vec{B}$ decreases with $1/B^2$.

Another interesting result from equation (3.21) is

$$\vec{i} \cdot \vec{E}_u = \frac{i^2}{\sigma} \tag{3.24}$$

Equation (3.24) shows that the component of $\vec{E}_u$ in the direction of $\vec{i}$ is independent of $\vec{B}$.

So far we consider only the case of fully ionized plasma. For a plasma of n species, e. g., partially ionized plasma, the condition is much more complicated. We should solve equation (3.19) for $\vec{w}_r$ and then obtain $\vec{i}$ by the help of equation (3.9). In general the final result may be written as follows (*4*):

$$\vec{i} = a \vec{E}_u + b \vec{E}_u \times \vec{B} + c (\vec{E}_u \times \vec{B}) \times \vec{B} \tag{3.25}$$

or

$$\vec{E}_u = R_{\parallel} \vec{i} + \zeta \vec{i} \times \vec{B} + (R_{\parallel} - R_{\perp}) \frac{(\vec{i} \times \vec{B}) \times \vec{B}}{B^2} \tag{3.26}$$

where $R_{\parallel} = \frac{1}{a}$, $\zeta = \frac{-b}{(a - c B^2)^2 + b^2 B^2}$, $R_{\perp} = \frac{a - c B^2}{(a - c B^2)^2 + b^2 B^2}$

$a = \frac{1}{R_{\parallel}}$, $b = \frac{-\zeta}{R_{\perp}{}^2 + \zeta^2 B^2}$, $c = \frac{1}{B^2} \cdot \left(\frac{1}{R_{\parallel}} - \frac{R_{\perp}}{R_{\perp}{}^2 + \zeta^2 B^2} \right)$

where a, b, and c are scalar functions of ρ_r, ρ_{er}, α_{rs}, and $|\vec{B}|$.

These notations were also first introduced by GRAD (*4*). If $\vec{E}_u$ is parallel to $\vec{B}$,

$$\vec{E}_u = R_{\parallel} \vec{i} \tag{3.27}$$

If $\vec{E}_u$ is perpendicular to $\vec{B}$, we have

$$\vec{E}_u = R_{\perp} \vec{i} + \zeta \vec{i} \times \vec{B} \tag{3.28}$$

In general $R_{\parallel}$ is not equal to $R_{\perp}$. Only when $R_{\parallel} = R_{\perp} = R$, equation (3.26) reduces to our simple relation (3.21).

If we take some of the gasdynamic forces into consideration, the electrical current equation will not be of the simple forms given above. In order to show some effects of gasdynamic forces on the electrical current, let us consider again the simple case of a fully ionized plasma. Now in addition to the electromagnetic forces, we shall include gasdynamic pressure gradient. If we assume that $T_1 = T_2 = T$, then in our case $p_1 = p_2 = p/2$. We obtain the following current equation instead of equation (3.21):

$$R\vec{i} + \zeta\vec{i} \times \vec{B} = \vec{E}_u - \frac{1}{\rho_{e1}} \nabla p_1 = \vec{E}_{ut} \tag{3.29}$$

The only difference between equations (3.21) and (3.29) is the additional term $-(1/\rho_{e1}) \nabla p_1$ which may be interpreted as additional electric field (*3*). The total effective electric field $\vec{E}_{ut}$ is then the sum of $\vec{E}_u$ and this additional field due to pressure gradient of the electrons.

In conclusion, the generalized Ohm's law is a highly simplified form of the electric current equation in plasma dynamics. For further research, it is highly desirable to use some more accurate relations of electrical current instead of the generalized Ohm's law. Since the development of plasma dynamics is still in its initial stage, at the present time we shall be content to use the generalized Ohm's law in our analysis of plasma dynamics, particularly for magnetogasdynamics in order to obtain some important results of the interaction of gasdynamics and electromagnetic fields.

5. Vector and scalar potentials (*2*). The divergence of equation (3.2) gives

$$\frac{\partial \nabla \cdot \mu_e \vec{H}}{\partial t} = 0$$

or $\nabla \cdot \mu_e \vec{H} =$ constant at every point in the field. This constant must be zero if ever in its past or future history the field $\vec{H}$ may vanish. Hence we have

$$\nabla \cdot \vec{B} = \nabla \cdot \mu_e \vec{H} = 0 \tag{3.30}$$

Similarly the divergence of equation (3.1) with the help of equation (3.8) gives

$$\frac{\partial}{\partial t}(\nabla \cdot \varepsilon \vec{E} - \rho_e) = 0$$

If in its past or future history the electric field $\vec{E}$ and ρ_e may vanish simultaneously, we have

$$\nabla \cdot \varepsilon \vec{E} = \rho_e \tag{3.31}$$

Under the conditions where ρ_e is negligible, we have

$$\nabla \cdot \varepsilon \vec{E} = 0 \tag{3.32}$$

Because of equation (3.30), we may introduce a vector potential $\vec{A}$ such that

$$\vec{B} = \nabla \times \vec{A} \tag{3.33}$$

or

$$\vec{H} = \frac{1}{\mu_e} \nabla \times \vec{A} \tag{3.33a}$$

with the condition

$$\nabla \cdot \vec{A} = 0 \tag{3.34}$$

Substituting equation (3.33) into equation (3.2) we gave

$$\nabla \times \left(\vec{E} + \frac{\partial \vec{A}}{\partial t}\right) = 0 \tag{3.35}$$

Because of equation (3.35) we may introduce a scalar potential such that

$$-\nabla \Phi = \vec{E} + \frac{\partial \vec{A}}{\partial t} \tag{3.36}$$

Sometimes it is more convenient to analyze the electromagnetic equations in terms of these vector and scalar potentials instead of the fields $\vec{H}$ and $\vec{E}$.

Substituting the above relations of these potentials into equation (3.1), we have (*2*)

$$\begin{aligned} \frac{1}{c^2} \cdot \frac{\partial^2 \vec{A}}{\partial t^2} - \nabla^2 \vec{A} + \mu_e \sigma \left[\frac{\partial \vec{A}}{\partial t} - \vec{u} \times (\nabla \times \vec{A})\right] = \\ = -\frac{1}{c^2}\left(\frac{\partial}{\partial t} \nabla \Phi + \vec{u}\, \nabla^2 \Phi\right) - \mu_e \sigma \nabla \Phi \end{aligned} \tag{3.37}$$

Equations (3.34) and (3.37) constitute a set of equations for the potentials $\vec{A}$ and Φ.

Under the conditions corresponding to equation (3.7), equation (3.37) becomes

$$\nabla^2 \vec{A} - \mu_e \sigma \left[\frac{\partial \vec{A}}{\partial t} - \vec{u} \times (\nabla \times \vec{A})\right] = \mu_e \sigma \nabla \Phi \tag{3.38}$$

6. Boundary conditions (*7*). The Maxwell's equations (3.1) and (3.2) or their equivalent equations are valid only for those points in whose neighborhood the physical properties of the medium vary continuously. On the boundary of the flow field, the physical properties of the medium may exhibit discontinuities. For instance, at a solid boundary, the electromagnetic properties of the plasma will change abruptly to those of the solid. Across such a surface of discontinuity of electromagnetic properties, the following four conditions hold (*7*):

(i) The transition of the normal component of magnetic induction $\vec{B} = \mu_e \vec{H}$ is continuous, i. e.,

$$(\vec{B}_2 - \vec{B}_1) \cdot \vec{n} = 0 \tag{3.39}$$

where $\vec{n}$ is the unit normal to the surface of discontinuity. Subscripts 1 and 2 refer to the values immediately on each side of the surface.

(ii) The behavior of the magnetic field $\vec{H}$ at this boundary is

$$\vec{n} \times (\vec{H}_2 - \vec{H}_1) = \vec{J}_s \tag{3.40}$$

where $\vec{J}_s$ is the surface current density. For finite electrical conductivity,

$\sigma \neq \infty$, $\vec{J}_s$ is zero; whereas for infinite electrical conductivity, $\sigma = \infty$, $\vec{J}_s$ may be different from zero.

(iii) The transition of the tangential component of the electric field $\vec{E}$ is continuous, i. e.,

$$\vec{n} \times (\vec{E}_2 - \vec{E}_1) = 0 \tag{3.41}$$

(iv) The behavior of dielectric displacement $\vec{D} = \varepsilon \vec{E}$ at this boundary is

$$\vec{n} \cdot (\vec{D}_2 - \vec{D}_1) = \rho_{es} \tag{3.42}$$

where ρ_{es} is the surface free charge density.

For most of our problems of magnetogasdynamics, we may neglect the surface current density $\vec{J}_s$ and the surface free charge density ρ_{es}. Hence our boundary conditions become that both the tangential components of $\vec{H}$ and $\vec{E}$, and the normal components of $\vec{B}$ and $\vec{D}$ are all continuous across a surface separating a body and a fluid or two fluids. The distinctions between $\vec{H}$ and $\vec{B}$ and between $\vec{E}$ and $\vec{D}$ should be noticed here because the values of μ_e and ε may be different on both sides of the boundary.

7. Units in electromagnetic theory (*6*). In the electromagnetic theory, besides the three fundamental dimensions of mechanics, e. g., length, mass, and time, we have another fundamental dimension, an electromagnetic unit. There are a number of different systems of units in electromagnetic theory in common use. This fact complicates the electromagnetic theory. It is found that the same equation, e. g. Maxwell's equation, has different forms in different papers or books. In order that the reader may transform our equations into a form with which he is familiar, we summarize briefly here those systems of unit commonly used in electromagnetic theory.

The most common systems of units used in electromagnetic theory are:

(i) The absolute cgs systems of electrostatic units in which the basic unit is usually abbreviated as esu.

(ii) The absolute cgs system of electromagnetic units in which the basic unit is usually abbreviated as emu.

(iii) Gaussian or mixed units which are a mixture of esu and emu.

(iv) Practical units which are largely based on emu but differ from emu by a power of ten in order to get convenient size.

(v) MKS or Giorgi system which is the one used in this book.

Furthermore in all these systems, we have a choice of "rationalized" and "unrationalized" units differing by factors of 4π.

The essential points of these systems are as follows:

(i) Unrationalized electrostatic units (esu).

In this system we take the centimeter as the unit of length, the gram as the unit of mass, and the second as the unit of time. The fundamental electrostatic unit is the unit charge which is defined by Coulomb's law:

$$F = \frac{QQ'}{r^2} \text{ dynes} \tag{3.43}$$

where F is the force between two charges Q and Q' and is expressed in dynes; r is the distance between these two charges in centimeters. The charges Q and Q' are expressed in unit charges. If Q nad Q' are unit charges and r is one centimeter, the force F will be one dyne.

The corresponding rationalized electrostatic unit charge would be defined by

$$F = \frac{Q_r\, Q_r'}{4\,\pi\, r^2} \text{ dynes} \tag{3.44}$$

Hence the unit charge in rationalized unit is $1/\sqrt{4\,\pi}$ times that in unrationalized units. We shall not consider the rationalized esu system any more.

The unit current is equal to unit charge passing any point of circuit per second.

The unit electric field $\vec{E}$ is defined by the equation:

$$\vec{F} = Q\,\vec{E} \text{ dynes} \quad (3.45$$

In a unit $\vec{E}$, the unit charge is acted on by a force of one dyne.

The unit of electric displacement $\vec{D}$ in the unrationalized esu system is determined by the arbitrary assumption that $\vec{D} = \vec{E}$ in free space, i. e.,

$$\vec{D} = \varepsilon\,\vec{E} = \vec{E} \tag{3.46}$$

Hence the induced capacity ε is taken as unity in free space.

The electrical conductivity σ is defined by Ohm's law:

$$\vec{J} = \sigma\,\vec{E} \tag{3.47}$$

where $\vec{J}$ is the current density in esu.

(ii) Unrationalized electromagnetic units (emu).

In this system, we still take cgs units for length, mass and time but the fundamental electric unit is the unit current in the electromagnetic system which is defined by the formula

$$\mathrm{d}\,\vec{F} = I_1\, I_2 \frac{\mathrm{d}\,\vec{s}_1 \times (\mathrm{d}\,\vec{s}_2 \times \vec{r})}{|\,r\,|^3} \text{ dynes} \tag{3.48}$$

where $I\,\mathrm{d}\,\vec{s}$ is the vector current element, I is the current in emu, and $\vec{r}$ is the spatial vector between the two current elements 1 and 2. The unit current in emu is the current which, flowing in a one-centimeter length of wire, acts on a similar current one centimeter away, appropriately oriented, with a force of one dyne.

The unit charge in emu is defined as the charge crossing a given point per second, when unit current is flowing.

The unit of electric field is still defined by equation (3.45) except that here both $\vec{E}$ and Q are in emu.

From experiments, it was found that the following relations between esu and emu of charge, current and electric field hold:

$$\begin{aligned}1 \text{ emu of charge} &= c \cdot 1 \text{ esu of charge}\\ 1 \text{ emu of current} &= c \cdot 1 \text{ esu of current}\\ c \cdot 1 \text{ emu of electric field} &= 1 \text{ esu of electric field}\end{aligned} \tag{3.49}$$

where c = velocity of light in free space $= 1/\sqrt{\varepsilon\,\mu_e} =$

$$= 2.99790 \cdot 10^{10} \text{ cm/sec.} \simeq 3 \cdot 10^{10} \text{ cm/sec.} \tag{3.50}$$

in which ε = inductive capacity in free space and μ_e = magnetic permeability in free space.

In emu, we define the magnetic induction $\vec{B}$ by

$$\vec{F} = Q\,(\vec{u} \times \vec{B}) \text{ dynes} \tag{3.51}$$

or

$$\frac{\vec{F}}{\text{volume}} = \vec{J} \times \vec{B} \text{ dynes/cm.} \tag{3.52}$$

where $\vec{J}$ is the current density, $\vec{u}$ is the velocity vector. Thus in the emu system the unit magnetic induction is that induction by which a unit current flowing in a wire one centimeter long at right angle to the magnetic field is acted on by a force of one dyne. This emu of magnetic induction is known as a gauss.

In emu, the magnetic field $\vec{H}$ in free space is defined as

$$\vec{H} = \vec{B} = \mu_e\,\vec{H} \tag{3.53}$$

where μ_e in free space is arbitrarily chosen as unity.

The electric displacement in free space is then

$$\vec{D} = \varepsilon\,\vec{E} = \frac{\vec{E}}{c^2} \tag{3.54}$$

In emu, Maxwell equations in free space may be written as follows:

$$\begin{aligned}&\nabla \times \vec{E} = -\frac{\partial \vec{H}}{\partial t}, \quad \nabla \cdot \vec{H} = 0\\ &\nabla \times \vec{H} = \frac{1}{c^2} \cdot \frac{\partial \vec{E}}{\partial t} + 4\pi \vec{J}, \quad \nabla \cdot \vec{E} = 4\pi c^2 \rho_e\end{aligned} \tag{3.55}$$

(iii) Gaussian or mixed units.

In this system we use emu for $\vec{B}$ and $\vec{H}$ and esu for all other quantities including current and charge. Maxwell's equations in free space in Gaussian units are then

$$\begin{aligned}&\nabla \times \vec{E} = -\frac{1}{c} \cdot \frac{\partial \vec{H}}{\partial t}, \quad \nabla \cdot \vec{H} = 0\\ &\nabla \times \vec{H} = \frac{1}{c^2} \cdot \frac{\partial E}{\partial t} + 4\pi \frac{\vec{J}}{c}, \quad \nabla \cdot \vec{E} = 4\pi c^3 \rho_e\end{aligned} \tag{3.56}$$

(iv) Practical units.

They are based on the emu but differ from emu by some arbitrary power of 10. Some of the important relations of practical units and emu are listed below:

Quantity	Practical unit	emu
current	1 ampere	$= 10^{-1}$ emu of current
potential	1 volt	$= 10^{8}$ emu of potential
work/second	1 ampere · 1 volt	= 1 watt = 10 erg/sec.
resistance	1 ohm	$= 10^{9}$ emu of resistance
magnetic induction ..	1 weber/cm.	$= 10^{8}$ gausses

The weber is unit of flux, not magnetic induction. Other practical units are:

$$1 \text{ ampere} = 1 \text{ coulomb/sec.}$$

A coulomb is a unit charge in practical units. The unit electric field in practical units is 1 volt/cm.

(v) MKS or Giorgi system.

Probably this is the most common unit system used in recent years. It is a rationalized MKS unit system in which the meter is used as the unit length; the kilogram, as the unit of mass; the second, as the unit of time and any electric quantity belonging to the practical unit system such as coulomb, ampere or ohm as the fourth unit. We shall use the coulomb as the fourth unit.

In this system, the magnetic permeability in free space has the magnitude

$$\mu_e = 4\pi \cdot 10^{-7} \frac{\text{kg} \cdot \text{m}}{(\text{coulomb})^2} \tag{3.57}$$

The value of ε in free space may be determined by equation (3.50), i. e.,

$$\varepsilon = 8.854 \cdot 10^{-12} \frac{\text{coulomb}^2 \, \text{sec}^2}{\text{kg} \cdot \text{m}^3} \tag{3.58}$$

Maxwell's equations in this system are then

$$\begin{aligned} \nabla \times \vec{E} &= -\frac{\partial \cdot \mu_e \vec{H}}{\partial t}, \quad \nabla \cdot \vec{H} = 0 \\ \nabla \times \vec{H} &= \vec{J} + \frac{\partial \varepsilon \vec{E}}{\partial t}, \quad \nabla \cdot \varepsilon \vec{E} = \rho_e \end{aligned} \tag{3.59}$$

References

1. BLANK, A. A., K. O. FRIEDRICHS, and H. GRAD: Theory of Maxwell's Equations without Displacement Current. Notes on Magneto-hydrodynamics, No. V. Inst. of Math. Sciences, New York Univ., Nov. 1, 1957.
2. BURGERS, J. M.: The Penetration of a Shock Wave into a Magnetic Field, Tech. Note No. BN-102, Inst. for Fluid Dynamics and Appl. Math. Univ. of Maryland, June 1957.
3. COWLING, T. G.: Magnetohydrodynamics, Interscience Publishers, Inc., New York, 1957.
4. GRAD, H.: Ohm's Law. Notes on Magneto-hydrodynamics, No. IV, Inst. of Math. Sciences, New York Univ., Aug. 1956.
5. PAI, S. I.: The Fundamental Equations of Magnetogasdynamics. Engineering report No. 10053-M, The Martin Company, Baltimore, April 1, 1958.
6. PAI, S. I.: Introduction to the Theory of Compressible Flow. Van Nostrand Co., Princeton, N. J., 1959.
7. STRATTON, J. A.: Electromagnetic Theory. McGraw-Hill Book Company, Inc., New York, 1941.

Chapter IV

Magnetogasdynamics and Electromagnetogasdynamics

1. Introduction. In this chapter, we consider the plasma as a single fluid. The fundamental equations are given in chapter II, § 8. They are as follows:

$$p = R_p \rho T \tag{4.1}$$

$$\frac{\partial \rho}{\partial t} + \frac{\partial}{\partial x^i} (\rho u^i) = 0 \tag{4.2}$$

$$\rho \frac{D u^i}{D t} = -\frac{\partial p_t}{\partial x^i} + \frac{\partial \tau^{ij}}{\partial x^j} + F_e{}^i + F_g{}^i \tag{4.3}$$

$$\frac{\partial \rho \overline{e_m}}{\partial t} + \frac{\partial \rho \overline{e_m} u^j}{\partial x^j} = -\frac{\partial u^j p_t}{\partial x^j} + \frac{\partial u^i \tau^{ij}}{\partial x^j} + E^j J^j + \frac{\partial Q^j}{\partial x^j} \tag{4.4}$$

$$\nabla \times \vec{H} = \vec{J} + \frac{\partial \varepsilon \vec{E}}{\partial t} \tag{4.5}$$

$$\nabla \times \vec{E} = -\frac{\partial \mu_e \vec{H}}{\partial t} \tag{4.6}$$

$$\frac{\partial \rho_e}{\partial t} + \frac{\partial J^j}{\partial x^j} = 0 \tag{4.7}$$

$$J^i = i^i + \rho_e u^i = \sigma [E^i + \mu_e (\vec{u} \times \vec{H})^i] + \rho_e u^i \tag{4.8}$$

$$\nabla \cdot \vec{H} = 0 \tag{4.9}$$

$$\nabla \cdot \vec{E} = \frac{1}{\varepsilon} \rho_e \tag{4.10}$$

where the unknowns to be investigated are u^i, p, ρ, T, E^i, H^i, ρ_e and J^i. Equations (4.9) and (4.10) are not independent equations which may be derived from equations (4.5), (4.6) and (4.7) (cf. chapter III, § 5); but they are important relations, hence we list them here for reference.

For many practically interesting problems such as astrophysical problems, we may simplify the above equations so that the problem becomes essentially the interaction of gasdynamic variables, $\vec{u}$, p, ρ, T and the magnetic field $\vec{H}$. The approximations used to simplify these equations may be called "Magnetogasdynamic approximations" which will be discussed in § 2. The corresponding fundamental equations of magnetogasdynamics will be given in § 3.

We shall discuss the important non-dimensional parameters in magnetogasdynamics in § 4. The magnetogasdynamic equations can be further simplified in various ranges of these non-dimensional parameters. The non-dimensional parameters of plasma dynamics, in general, will be discussed in chapter V where the relations of plasma dynamics and magnetogasdynamics from the point of view of these parameters will also be discussed.

Some of the properties of the fundamental equations of magnetogasdynamics will be discussed in § 5 and in chapters VI to XII.

Strictly speaking magnetogasdynamics holds good for the case of very large electrical conductivity, i. e., $\sigma \to \infty$ because in this case, the energy in the electric field under magnetogasdynamic approximations is negligible in comparison with that in the magnetic field. It may not be true when the electrical conductivity is very small, particularly in laboratory experiments where a large external electric field may be applied. In the latter case, we should consider the electric field and the magnetic field simultaneously as independent variables. The resultant analysis should be called "electromagnetogasdynamics." We shall discuss the fundamental equations of electromagnetogasdynamics in § 6. We shall also discuss several other cases which are different from the ordinary magnetogasdynamics.

2. Magnetogasdynamic approximations. For many practically interesting problems, the following conditions are satisfied:

(i) The time scale t_0 of our problem is of the same order of magnitude as L/U where L is the characteristic length and U is the characteristic velocity of the flow field. In other words, the non-dimensional parameter

$$R_t = \frac{t_0 U}{L} \tag{4.11}$$

is of the order of unity. It means that we shall not consider phenomena of very high frequency.

For example, if we take $U = 10^4$ m/sec. which is a typical high speed of a satellite and $L = 1$ m., we have $L/U = 10^{-4}$ sec. We shall not consider any high frequency phenomena with frequency much larger than $10^4\ \text{sec}^{-1}$.

(ii) The electrical field which may be characterized by a value E_0 is of the same order of magnitude as the induced electric field $\mu_e \vec{u} \times \vec{H}$. In other words, the non-dimensional parameter

$$R_E = \frac{E_0}{\mu_e U H_0} \tag{4.12}$$

is of the order of unity or smaller where H_0 is the characteristic magnetic field strength. This is a good assumption for very large electric conductivity σ, because as $\sigma \to \infty$, we would expect that

$$E \cong -\mu_e \vec{u} \times \vec{H} \tag{4.13}$$

otherwise the conduction current $\vec{i}$ will become very large for a slight motion of the plasma.

(iii) The velocity of the flow of a plasma is much smaller than the velocity of light c, i. e.,

$$R_c = \frac{U^2}{c^2} = U^2 \mu_e \varepsilon \ll 1 \tag{4.14}$$

For instance, if $U = 10^4$ m/sec., R_c is only $1.11 \cdot 10^{-9}$.

(iv) The radiation terms p_R and E_R are negligible.

(v) The non-electric force $F_g{}^i$ is also negligible.

From the first three approximations, it may be shown that the displacement current $\partial\, \varepsilon\, \vec{E}/\partial\, t$ and the excess electric charge ρ_e are negligible in our fundamental equations of plasma and that the energy in the electric field is much smaller than that in the magnetic field. As a result, all the electromagnetic variables may be expressed in terms of magnetic field. We study then the interaction of the magnetic field and the gasdynamic field only. Hence the term of "magnetogasdynamics" is introduced.

The approximations (iv) and (v) are not directly connected to the magnetogasdynamic; but they are assumed implicitly in ordinary investigations of magnetogasdynamics, so that our fundamental equations of plasma dynamics may be reduced to the exact form of conventional fundamental equations of magnetogasdynamics.

We introduce the following non-dimensional quantities:

$$\begin{gathered} x^* = \frac{x}{L}, \quad \nabla^* = \nabla \cdot L, \quad t^* = \frac{t}{t_0}, \quad \vec{u}^* = \frac{\vec{u}}{U} \\ \vec{E}^* = \frac{\vec{E}}{E_0}, \quad \vec{H}^* = \frac{\vec{H}}{H_0}, \quad \vec{J}^* = \frac{\vec{J}}{\sigma\, \mu_e\, U\, H_0} \end{gathered} \tag{4.15}$$

The Maxwell's equation (4.5) in the non-dimensional variables is

$$\frac{1}{R_\sigma} \nabla^* \times \vec{H}^* = \vec{J}^* + \frac{R_c R_E}{R_\sigma R_t} \cdot \frac{\partial \vec{E}^*}{\partial t^*} \tag{4.16}$$

$$\text{where } \mathrm{R}_\sigma = \sigma\, \mu_e\, UL = \frac{U\,L}{\nu_H} = \text{magnetic Reynolds number} \tag{4.17}$$

$$\text{and } \nu_H = \frac{1}{\sigma\, \mu_e} = \text{magnetic viscosity} = \text{magnetic diffusivity} \tag{4.18}$$

It should be noticed that all the non-dimensional variables defined in equation (4.15) are of the order of magnitude of unity. From equation (4.16) we see that under magnetogasdynamic approximations the displacement current is negligibly small in comparison with the curl of magnetic field. Hence equation (4.5) in magnetogasdynamics becomes [cf. equation (3.7a)]

$$\nabla \times \vec{H} = \vec{J} \tag{4.19}$$

In terms of the non-dimensional variables, the electric current equation (4.8) becomes

$$\vec{J}^* = (R_E\, \vec{E}^* + \vec{u}^* \times \vec{H}^*) + \frac{R_c\, R_E}{R_\sigma} (\nabla^* \cdot \vec{E}^*)\, \vec{u}^* \tag{4.20}$$

If $R_c/R_\sigma \ll 1$, the convection current $\rho_e\, \vec{u}$ is negligible in comparison with the conduction current $\vec{i}$ in magnetogasdynamics. For a conducting fluid, it is reasonable to assume that $R_c/R_\sigma \ll 1$. For instance, if we take $\sigma = 1$ mho/cm. $=$ $= 100$ mho/m. which is a typical value for an ionized air, $R_\sigma = 4\pi \cdot 10^{-5}\, UL$. Now if $U = 10^4$ m/sec. and $L = 1$ m., $R_\sigma = 1.264$ and $R_c/R_\sigma = 8.75 \cdot 10^{-8} \ll 1$.

When $R_c/R_\sigma \ll 1$, we have

$$\vec{J} = \vec{i} = \sigma\,(\vec{E} + \mu_e\,\vec{u} \times \vec{H}) \tag{4.21}$$

or

$$\vec{E} = \frac{\vec{J}}{\sigma} - \mu_e\,\vec{u} \times \vec{H} = \frac{1}{\sigma}(\nabla \times \vec{H}) - \mu_e\,\vec{u} \times \vec{H} \tag{4.22}$$

By equations (4.19) and (4.22) and the approximation $\rho_e = 0$, we see that all the electromagnetic variables may be expressed in terms of magnetic field $\vec{H}$ and then we study only the interaction of magnetic field with the gasdynamic field.

The electric energy is $\varepsilon\, E^2/2$ and the magnetic energy is $\mu_e\, H^2/2$. Under magnetogasdynamic approximations, the order of magnitude of these two energies is as follows:

$$\frac{\varepsilon\, E^2}{2} \sim \frac{\varepsilon\, E_0^2}{2} \sim \frac{\varepsilon\, \mu_e^2\, U^2\, H_0^2}{2} = R_c\,\frac{\mu_e\, H_0^2}{2} \sim R_c\,\frac{\mu_e\, H^2}{2} \tag{4.23}$$

Since $R_c \ll 1$, the electric energy is much smaller than the magnetic energy.

3. Fundamental equations of magnetogasdynamics (7). In magnetogasdynamics, our unknowns are H^i, u^i, p, ρ and T. We need two vector equations and three scalar equations for these unknowns. Substituting equations (4.19) and (4.22) into equation (4.6), we have the vector equation for the magnetic field $\vec{H}$, i. e.,

$$\frac{\partial \vec{H}}{\partial t} = \nabla \times (\vec{u} \times \vec{H}) - \nabla \times [\nu_H\,(\nabla \times \vec{H})] \tag{4.24}$$

Under the magnetogasdynamic approximations, the equation of motion (4.3) may be written as follows:

$$\rho\,\frac{D\vec{u}}{Dt} - \mu_e\,(\vec{H} \cdot \nabla)\,\vec{H} = -\nabla\left(p + \frac{\mu_e\, H^2}{2}\right) + (\nabla \cdot \tau) \tag{4.25}$$

where τ is the stress tensor and $(\nabla \cdot \tau)^i = \partial\, \tau^{ij}/\partial\, x^j$.

Similarly the energy equation (4.4) may be written as

$$\rho\,\frac{D\,h_0}{D\,t} = \frac{\partial\, p}{\partial\, t} + \nabla \cdot (\vec{u} \cdot \tau) + \nabla \cdot (\varkappa\, \nabla\, T) + (\nabla \times \vec{H}) \cdot [\mu_e\, \nu_H\,(\nabla \times \vec{H}) - \mu_e\, \vec{u} \times \vec{H}] \tag{4.26}$$

where $h_0 = C_p\, T + u^2/2 =$ stagnation enthalpy and C_p is the specific heat at constant pressure.

The fundamental equations of magnetogasdynamics are then equations (4.1), (4.2), (4.24), (4.25) and (4.26).

We are going to rewrite these fundamental equations in non-dimensional form. Besides the non-dimensional quantities of (4.15), we introduce the following non-dimensional quantities.

$$p^* = \frac{p}{\rho_0 U^2}, \quad \rho^* = \frac{\rho}{\rho_0}, \quad T^* = \frac{T}{T_0}, \quad \tau^* = \left(\mu_0 \frac{U}{L}\right)^{-1} \tau$$
$$\mu^* = \frac{\mu}{\mu_0}, \quad \varkappa^* = \frac{\varkappa}{\varkappa_0}, \quad \nu_H^* = \frac{\nu_H}{\nu_{H_0}}, \quad C_p^* = \frac{C_p}{C_{p0}} \tag{4.27}$$

where subscript 0 refers to some reference value.

The fundamental equations of magnetogasdynamics in terms of these non-dimensional quantities with $R_t = 1$, are as follows:

(i) Equation of state (4.1)

$$\gamma M_0^2 p^* = \rho^* T^* \tag{4.28}$$

(ii) Equation of continuity (4.2)

$$\frac{\partial \rho^*}{\partial t^*} + \frac{\partial}{\partial x^{i*}} (\rho^* u^{i*}) = 0 \tag{4.29}$$

(iii) Equation of magnetic field (4.24)

$$\frac{\partial \vec{H}^*}{\partial t^*} = \nabla^* \times (\vec{u}^* \times \vec{H}^*) - \frac{1}{R_\sigma} \{\nabla^* \times [\nu_H^* (\nabla^* \times \vec{H}^*)]\} \tag{4.30}$$

(iv) Equation of motion (4.25)

$$\rho^* \frac{D\vec{u}^*}{Dt^*} - R_H (\vec{H}^* \cdot \nabla^*) \vec{H}^* = -\nabla^* \left(p^* + R_H \frac{H^{*2}}{2}\right) + \frac{1}{R_e} (\nabla^* \cdot \tau^*) \tag{4.31}$$

(v) Equation of energy (4.26)

$$\rho^* \frac{D}{Dt^*} \left[C_p^* T^* + (\gamma - 1) M_0^2 \frac{u^{*2}}{2}\right] =$$
$$= (\gamma - 1) M_0^2 \frac{\partial p^*}{\partial t^*} + \frac{(\gamma - 1) M_0^2}{R_e} \nabla^* \cdot (\vec{u}^* \cdot \tau^*) + \frac{1}{P_r R_e} \nabla^* \cdot (\varkappa^* \nabla^* T^*)$$
$$+ (\gamma - 1) M_0^2 R_H (\nabla^* \times \vec{H}^*) \cdot \left[\frac{\nu_H^*}{R_\sigma} (\nabla^* \times \vec{H}^*) - (\vec{u}^* \times \vec{H}^*)\right] \tag{4.32}$$

4. Important parameters in magnetogasdynamics. From our non-dimensonial fundamental equations of magnetogasdynamics (4.28) to (4.32), we find that the following non-dimensional parameters characterize our problem:

$$\gamma = \text{ratio of specific heats} = \frac{C_{p0}}{C_{v0}} \tag{4.33}$$

$$M_0 = \text{Mach number} = \frac{U}{\sqrt{\gamma R_p T_0}} = \frac{U}{a_0} \tag{4.34}$$

$$R_e = \text{Reynolds number} = \frac{\rho_0 U L}{\mu_0} = \frac{U L}{\nu_0} \tag{4.35}$$

$$P_r = \text{Prandtl number} = \frac{C_{p0} \mu_0}{\varkappa_0} \tag{4.36}$$

$$R_H = \text{magnetic pressure number} = \frac{\mu_e H_0^2}{\rho_0 U^2} = \frac{V_H^2}{U^2} \tag{4.37}$$

$$R_\sigma = \text{magnetic Reynolds number} = \frac{U\,L}{\nu_{H_e}} = \frac{L}{L_e} = \frac{U}{V_e} \tag{4.38}$$

The parameters γ, M_0, R_e and P_r are the well known parameters in gasdynamics while R_H and R_σ are two new magnetogasdynamic parameters[1]. The significance of these parameters are as follows (*8*).

(a) The ratio of specific heats. The ratio of specific heats γ is a measure of the relative internal complexity of the molecules of the plasma.

(b) Mach number M_0. It is a measure of the compressibility of the fluid due to a high flow velocity and is defined as the ratio of flow velocity U to the sound speed a_0. It is easy to show that the ratio of the variation of density of the fluid to the variation of velocity is, to the first approximation, proportional to the square of the Mach number of the flow. Hence for very small Mach number, the variation of density, i. e., the compressibility effect, due to the variation of velocity of the flow field is negligibly small and the fluid may be considered as incompressible. For large Mach number, the effect of compressibility must be considered.

(c) Reynolds number R_e. It is a measure of the ratio of inertial force to the viscous force. When the Reynolds number of a system is small, the viscous force is predominant and the effect of viscosity is important in the whole flow field. When the Reynolds number is large, the inertial force is predominant and the effect of viscosity is important only in the narrow boundary layer region near the solid boundary or in any other region of large variation in velocity such as inside of a shock wave.

(d) Prandtl number P_r. It is a measure of the relative importance of viscosity and heat conduction. The Prandtl number may be written as follows:

$$P_r = \frac{\nu_0}{\dfrac{\varkappa_0}{C_{p0}\,\rho_0}} = \frac{\text{kinematic viscosity}}{\text{thermal diffusivity}} \tag{4.39}$$

The value of ν_0 shows the effect of viscosity of a fluid. If other things are the same, the smaller the value of ν_0, the narrower will be the region affected by viscosity. This region is known as the boundary layer region when ν_0 is very small. While ν_0 shows the momentum diffusivity due to viscosity effect, $\varkappa_0/C_{p0}\,\rho_0$ shows the thermal diffusivity due to heat conduction. The smaller the value of $\varkappa_0/C_{p0}\,\rho_0$, the narrower will be the region affected by heat conduction. This region is known as thermal boundary layer when $\varkappa_0/C_{p0}\,\rho_0$ is very small. Thus the Prandtl number shows the relative importance of viscosity and heat conduction of a fluid.

(e) Magnetic pressure number R_H. It is the ratio of the magnetic pressure $\mu_e H_0{}^2/2$ over the dynamic pressure $\rho_0\,U^2/2$. Only when R_H is of the order of unity or larger, the fluid flow will be affected noticeably by the magnetic field. If R_H is much smaller than unity, the terms due to magnetic field in both the equation of motion and that of energy may be neglected, and the gasdynamic motion will not be affected noticeably by the magnetic field.

---

[1] The typical values of these parameters are given in chapter V.

The velocity

$$V_H = H_0 \sqrt{\frac{\mu_e}{\rho_0}} \tag{4.40}$$

is known as the characteristic velocity of Alfvén's wave (cf. chapter X). It was ALFVÉN (*1*) who first showed that if there is a homogeneous magnetic field $\vec{H}_0$ in an incompressible and inviscid fluid of density ρ_0 and infinite electrical conductivity $\sigma_0 = \infty$, the disturbance in this liquid will propagate as a wave in the direction of $\vec{H}_0$ with a speed of V_H. In a compressible fluid, the magnetogasdynamic wave is much more complicated than the simple Alfvén's wave, but such waves have close relation with Alfvén's wave because the velocity V_H is still the characteristic velocity in the fundamental equations of magnetogasdynamics (cf. chapter X).

(f) Magnetic Reynolds number R_σ. This number has a similar form as ordinary Reynolds number with the magnetic viscosity ν_{H_0} in place of kinematic viscosity ν_0. It is the reason that R_σ is usually called Magnetic Reynolds number. However there is a big difference in the physical significance between R_e and R_σ. Ordinary Reynolds number R_e is obtained from dynamic consideration as the ratio of inertial force to the viscous force while magnetic Reynolds number is obtained from kinematic consideration as how the magnetic field will be influenced by the flow motion. R_σ may be regarded either as the ratio of the dimension of the flow field L to the characteristic length L_e, where

$$L_e = \frac{1}{\sigma_0 \mu_e U} = \frac{\nu_{H_0}}{U} \tag{4.41}$$

or as the ratio of the flow velocity U to a characteristic velocity V_e, where

$$V_e = \frac{1}{\sigma_0 \mu_e L} = \frac{\nu_{H_0}}{L} \tag{4.42}$$

The length L_e may be regarded as a characteristic length of which the magnetic field is moving through a conductor. It is the well known characteristic length in the skin effect of ordinary electromagnetic theory (*10*). If $L \gg L_e$, i. e., $R_\sigma \gg 1$, the magnetic field will stay with the flow, the so-called frozen-in fields, and it will be greatly influenced by the motion of the fluid. On the other hand if $L \ll L_e$, i. e., $R_\sigma \ll 1$, the magnetic field will not be influenced noticeably by the motion of the fluid.

The velocity V_e may be regarded as a characteristic velocity of which the magnetic field is moving through the conductor. If $U \gg V_e$, i e., $R_\sigma \gg 1$, the field is practically compelled to follow the motion, frozen-in field and it will be greatly influenced by the motion. On the other hand, if $U \ll V_e$, i. e., $R_\sigma \ll 1$, the magnetic field will not be influenced noticeably by the motion.

It should be noticed that even though the parameters γ, M_0, R_e, P_r, R_H and R_σ are sufficient to describe the magnetogasdynamic motion, they are not unique. One may obtain other important parameters from the combination of these six parameters. These new parameters may be more appropriate for special problems. For instance, the following two parameters have been used by many authors:

(g) Hartmann number R_h. It is obtained from the ratio of the magnetic force to the viscous force and is defined as

$$R_h = \sqrt{R_e R_H R_\sigma} = \mu_e H_0 L \left(\frac{\sigma_0}{\mu_0}\right)^{\frac{1}{2}} = \sqrt{\frac{\sigma_0 \mu_e^2 H_0^2 U}{\mu_0 \frac{U}{L^2}}} = \sqrt{\frac{\text{magnetic force}}{\text{viscous force}}} \tag{4.43}$$

It was first used by HARTMANN (*4*) in the study of flow in channels where the important forces are the magnetic force and viscous force.

(h) Magnetic parameter R_M. It is obtained from the ratio of the magnetic force to the inertial force and is defined as

$$R_M = \sqrt{R_H R_\sigma} = \mu_e H_0 \left(\frac{\sigma_0 L}{\rho_0 U}\right)^{\frac{1}{2}} = \sqrt{\frac{\sigma_0 \mu_e^2 H_0^2 U}{\rho_0 \frac{U^2}{L}}} = \sqrt{\frac{\text{magnetic force}}{\text{inertial force}}} \tag{4.44}$$

Our fundamental equations (4.28) to (4.32) may be further simplified in certain special range of values of the above parameters. We shall list a few important cases below:

(i) Case of very small R_σ. If the value of R_σ is very small, we see that the first term on the right hand side of equation (4.30) will be negligibly small in comparison to the second term which is proportional to $1/R_\sigma$. In this case the magnetic field $\vec{H}$ is practically independent of the flow motion. Then we may assume that $\vec{H}$ is given and solve the gasdynamic equations for u^i, p, ρ and T. In this case the fundamental equations are

$$p = R_p \rho T \tag{4.45a}$$

$$\frac{\partial \rho}{\partial t} + \nabla \cdot (\rho \vec{u}) = 0 \tag{4.45b}$$

$$\rho \frac{D\vec{u}}{Dt} - \mu_e (\vec{H} \cdot \nabla) \vec{H} = -\nabla \left(p + \frac{\mu_e H^2}{2}\right) + (\nabla \cdot \tau) \tag{4.45c}$$

$$\rho \frac{D h_0}{Dt} = \frac{\partial p}{\partial t} + \nabla \cdot (\vec{u} \cdot \tau) + \nabla \cdot (\varkappa \nabla T) + (\nabla \times \vec{H}) \cdot [\mu_e \nu_H (\nabla \times \vec{H}) - \mu_e \vec{u} \times \vec{H}] \tag{4.45d}$$

where $\vec{H} = \vec{H}_0$ is the given external magnetic field. We neglect the induced magnetic field here. This is a rather important case for some practical engineering problem where the conductivity is not large but the strength of magnetic field $\vec{H}_0$ is large. Equation (4.45) gives the first order approximation how the flow field may be affected by the magnetic field.

(ii) Case of very small R_H. If the value of R_H is very small, the gasdynamic equations (4.28), (4.29), (4.31), and (4.32) are practically independent of the magnetic field. We may solve the flow problem from the pure gasdynamic equations:

$$p = R_p \rho T \tag{4.46a}$$

$$\frac{\partial \rho}{\partial t} + \nabla \cdot (\rho \vec{u}) = 0 \tag{4.46b}$$

$$\rho \frac{D\vec{u}}{Dt} = -\nabla p + (\nabla \cdot \tau) \tag{4.46c}$$

$$\rho \frac{D h_0}{Dt} = \frac{\partial p}{\partial t} + \nabla \cdot (\vec{u} \cdot \tau) + \nabla \cdot (\varkappa \nabla T) \tag{4.46d}$$

for $\vec{u}$, p, ρ, and T. After these variables are known, we may solve equation (4.24) for $\vec{H}$. It should be noticed that in general ν_H is a function of T. Here we study the variation of magnetic field under the influence of flow field.

(iii) Low Mach number case and the incompressible fluid case. If the Mach number M_0 is very small or zero, the equation of state (4.1) may be replaced by ρ = constant = ρ_0. Our unknowns are then H^i, u^i, p and T. These unknowns are divided into two groups: one group consists of H^i, u^i and p and the other, all variables including T. The fundamental equations for the group of H^i, u^i and p are

$$\nabla \cdot \vec{u} = 0 \tag{4.47a}$$

$$\rho \frac{D\vec{u}}{Dt} - \mu_e (\vec{H} \cdot \nabla) \vec{H} = -\nabla \left(p + \frac{\mu_e H^2}{2}\right) + \mu \nabla^2 \vec{u} \tag{4.47b}$$

$$\frac{\partial \vec{H}}{\partial t} + (\vec{u} \cdot \nabla) \vec{H} - (\vec{H} \cdot \nabla) \vec{u} = \nu_H \nabla^2 \vec{H} \tag{4.47c}$$

where we assume that both μ and ν_H are constant.

From the energy equation (4.32), if $M_0 \to 0$ but $M_0^2 R_H$ and $M_0^2 R_H/R_\sigma$ are not negligibly small, we have the equation for T as follows:

$$\rho \frac{D C_p T}{Dt} = \nabla \cdot (\varkappa \nabla T) + (\nabla \times \vec{H}) \cdot [\mu_e \nu_H (\nabla \times \vec{H}) - \mu_e \vec{u} \times \vec{H}] \tag{4.48}$$

In equation (4.48), we assume that the joule heat and work due to magnetic force are of the same order of magnitude as the thermal energy, because the velocity U does not enter in the product of $M_0^2 R_H$. For small value of M_0, the product $M_0^2 R_H$ may not be small.

If both $M_0^2 R_H$ and $M_0^2 R_H/R_\sigma$ are negligibly small, equation (4.48) becomes

$$\rho \frac{D C_p T}{Dt} = \nabla \cdot (\varkappa \nabla T) \tag{4.48a}$$

This is the thermal equation of ordinary gasdynamics at very low Mach number.

For an incompressible fluid, if $M_0^2 R_H/R_\sigma$ is not small, the thermal equation (4.48a) should be replaced by the following equation:

$$\rho \frac{D C_v T}{Dt} = \nabla \cdot (\varkappa \nabla T) + \mu_e \nu_H (\nabla \times \vec{H})^2 \tag{4.49}$$

If $M_0^2 R_H/R_\sigma$ is also small, equation (4.49) reduces to

$$\rho \frac{D C_v T}{Dt} = \nabla \cdot (\varkappa \nabla T) \tag{4.49a}$$

This is the thermal equation for incompressible fluid.

Our problem in the present case is first to solve for $\vec{H}$, $\vec{u}$, and p from equations (4.47) and then to solve for T with the known values of $\vec{H}$ and $\vec{u}$ from equation (4.48) or (4.49). In solving equations (4.47) we may also treat those special cases of small R_σ and small R_H in the same manner discussed in (i) and (ii) respectively.

(iv) Ideal plasma case. An inviscid, non-heat-conducting and infinitely electrically conducting plasma may be called an ideal plasma. In this case, we have $R_e = \infty$, $R_\sigma = \infty$ and P_r = finite. The dissipation terms in equation (4.32), i. e., the terms of heat conduction, viscous dissipation and joule heat, are zero. The energy equation becomes simply

$$\frac{DS}{Dt} = 0 \tag{4.50a}$$

where S is the entropy of the plasma. Since $p = p(\rho, S)$, we may consider the unknowns in our problem to be $\vec{H}$, $\vec{u}$, ρ and S. In addition to equation (4.50a), the fundamental equations for this ideal plasma are

$$\frac{\partial \rho}{\partial t} + \nabla \cdot (\rho \vec{u}) = 0 \tag{4.50b}$$

$$\rho \frac{D\vec{u}}{Dt} - \mu_e (\vec{H} \cdot \nabla) \vec{H} = -\nabla \left(p + \frac{\mu_e H^2}{2} \right) \tag{4.50c}$$

$$\frac{\partial \vec{H}}{\partial t} = \nabla \times (\vec{u} \times \vec{H}) \tag{4.50d}$$

Equation (4.50) is known as LUNDQUIST (*6*) equation and has been studied extensively by FRIEDRICHS (*3*) (cf. chapter XII).

5. Some properties of the fundamental equation of magnetogasdynamics. The fundamental equations of magnetogasdynamics (4.28) to (4.32) are nonlinear and of high order because of the coefficient of viscosity, heat conduction and electrical conductivity are all in general functions of temperature and thus functions of velocity. This system of equations is more complicated than the fundamental equations of viscous compressible fluids which are sometimes called the Navier-Stokes equations of a compressible fluid, and whose properties have not been completely known yet. We would expect that we know even less of the mathematical properties of the fundamental equations of magnetogasdynamics. However there are many similarities between the Navier-Stokes equations and the magnetogasdynamic equations. As a result, we may apply those methods used successfully in ordinary gasdynamics to treat the problems of magnetogasdynamics, as we shall see in the following chapters. Here we are going to show two well-known similarities between magnetogasdynamics and ordinary gasdynamics.

(a) Analogy of magnetic field of magnetogasdynamics and vorticity in ordinary gasdynamics. If we assume that the electrical conductivity σ is constant, the equations which govern the magnetic field $\vec{H}$ in magnetogasdynamics are

$$\nabla \cdot \vec{H} = 0 \tag{4.51a}$$

$$\frac{\partial \vec{H}}{\partial t} + (\vec{u}\cdot\nabla)\vec{H} - (\vec{H}\cdot\nabla)\vec{u} + \vec{H}(\nabla\cdot\vec{u}) = \nu_H \nabla^2 \vec{H} \tag{4.51b}$$

[cf. equations (4.9) and (4.24)]. Equations (4.51) are formally identical to the equations which govern the vorticity $\vec{\omega} = \nabla \times \vec{u}$ in ordinary gasdynamics if we assume that the coefficient of viscosity μ is constant and that the fluid is barotropic, i. e., the pressure is a function of density only. The equations which govern the vorticity $\vec{\omega}$ under these assumptions are (7):

$$\nabla \cdot \vec{\omega} = 0 \tag{4.52a}$$

$$\frac{\partial \vec{\omega}}{\partial t} + (\vec{u}\cdot\nabla)\vec{\omega} - (\vec{\omega}\cdot\nabla)\vec{u} + \vec{\omega}(\nabla\cdot\vec{u}) = \nu \nabla^2 \vec{\omega} \tag{4.52b}$$

Since equations (4.51) and (4.52) are identical in form, we may apply all the well known theorems of vorticity in barotropic fluid of ordinary gasdynamics to the magnetic field of magnetogasdynamics. Particularly, by Helmholtz's theorem we know that the vortex lines move with the fluid if $\nu = 0$; hence we may also show that the lines of magnetic force move with the fluid if $\nu_H = 0$, i. e., when $\sigma = \infty$.

(b) Ideal plasma. The system of fundamental equations of an ideal plasma (4.50) is hyperbolic and it is similar to that of ordinary dynamics for an inviscid fluid. It is well known that for hyperbolic equations, disturbances are propagated with finite speed. For an inviscid fluid in ordinary gasdynamics, disturbances travel with the speed of sound relative to the motion of the fluid. In magnetogasdynamics of ideal plasma, we would expect that the disturbances also travel with some finite speeds. Since the equations of magnetogasdynamics are more complicated than those of ordinary gasdynamics, we would expect that the modes of disturbances propagation in magnetogasdynamics are more complicated than those in ordinary gasdynamics. We shall discuss these in chapter XII. One of the most powerful methods used to solve hyperbolic equation is the method of characteristics which has been used very successfully in ordinary gasdynamics. In chapter XII we shall also show that the method of characteristics is also a powerful method in treating the problems of ideal plasma.

6. Electromagnetogasdynamics. From above discussions, we see that magnetogasdynamics (4.28) to (4.32) is a special branch of the more general plasma dynamics (4.1) to (4.8). Even though a large class of problems satisfies the magnetogasdynamic approximations of § 2, particularly many astrophysical and geophysical problems, we may relax some of the conditions of magnetogasdynamic approximations in order to improve the range of validity of our simplified fundamental equations or their modifications. We discuss the magnetogasdynamic approximation further and show various cases which give more general results than those given by the restricted magnetogasdynamics (4.28) to (4.32).

(a) Relativistic effects. The condition (4.14), i. e., $R_c \ll 1$, may be considered as always satisfied in ordinary flow problems. If it is not satisfied, the relativistic effect (5) should be considered. In that case, even the plasma dynamic equations (4.1) to (4.8) are no longer accurate. One should modify the fundamental equations by taking the relativistic effects into account, i. e., the mass should be considered as a function of velocity. It is beyond the scope of this book.

(b) High frequency phenomena. The condition (4.11), i. e., $R_t = 0\,(1)$ or larger, may be considered as satisfied for most flow problems because the very high frequency phenomena are not important in flow problems which are mostly slowly unsteady flow or even steady cases. For the latter case the high frequency phenomena do not enter at all. However if one considers such problems as plasma oscillations, the high frequency phenomena may not be ignored. In that case, we should discard the approximation $R_t = 0\,(1)$ and retain the displacement current in the Maxwell's equation. It is of interest to notice that when $R_t = 0\,(1)$ and the displacement current may be neglected, the light velocity will not be a characteristic velocity in the treatment of wave motion in magnetogasdynamics. We shall discuss various wave motions with and without displacement current in chapter X.

(c) Electromagnetogasdynamics. With the magnetogasdynamic approximations, $R_c \ll 1$, $R_t = 0\,(1)$ and $R_E = 0\,(1)$, we have shown in equation (4.23) that the energy in the electric field is much smaller than that in the magnetic field. Under these conditions, we usually neglect the external electric field and consider only the influence of external magnetic field on the phenomena. We obtain the name of magnetogasdynamics. Of course in magnetogasdynamics, both the induced electric field and induced magnetic field are considered. Even though the neglect of external electric field is a good approximation in cosmical problems, it may not be a good approximation in the treatment of problems of laboratory experiments. In laboratory experiments, the current circuit must be closed. Usually the current circuit is closed through some resistance outside the flow field investigated. If the resistance is negligibly small, we have the case of magnetogasdynamics. If the resistance is not negligibly small, an electric potential difference will be set up which would affect the electric and the magnetic fields in the flow field investigated. We should take this external electric field into consideration in our analysis. We may also put an arbitrary external electric field $\vec{E}_0$ in our experiment so that the energy in the electric field may be of the same order of magnitude as that in the magnetic field. The question is then what modifications should be made on our fundamental equations of magnetogasdynamics so that the case that of the electric energy not being negligibly small may also be studied. We are going to reexamine our magnetogasdynamic approximations with the exception that R_E may be much larger than unity.

From equation (4.16), we see that if $R_c\, R_E/R_t \ll 1$, the displacement current is negligible. If $R_t = 0\,(1)$ or larger, we conclude that if $R_c\, R_E \ll 1$, the displacement current is negligible. From equation (4.20), we see that if $R_c\, R_E/R_\sigma \ll 1$, the convection current is negligible. If $R_\sigma = 0\,(1)$ or larger, we conclude that if $R_c\, R_E \ll 1$, the convection current is negligible. Equation (4.23) may be rewritten as follows:

$$\frac{\varepsilon E^2}{2} \sim \frac{\varepsilon E_0{}^2}{2} = R_E{}^2 \frac{\varepsilon \mu_e{}^2 U^2 H_0{}^2}{2} = R_E{}^2 R_c \frac{\mu_e H_0{}^2}{2} \sim R_E{}^2 R_c \frac{\mu_e H^2}{2} \tag{4.53}$$

The ratio of the energy in electric field to that in the magnetic field is of the order of $R_E{}^2 R_c$. Since R_c is usually very small, it is possible

$$R_E{}^2 R_c = 0\,(1) \quad \text{and} \quad R_E\, R_c \ll 1 \tag{4.54}$$

Under the conditions of (4.54), with $R_t = 0$ (1) and $R_\sigma = 0$ (1) or larger, we see that the energies in the electric field and the magnetic field are of the same order of magnitude but the displacement current and convection current are still negligible. As a result, our fundamental equations in the present case are the same as those of magnetogasdynamics. However in the present case, we should take both the external electric field and external magnetic field into consideration in our analysis of the flow problem. We should call the present problem as "electromagnetogasdynamics." In treatment of electromagnetogasdynamics, it is more convenient to retain the variables $\vec{E}$ and $\vec{J}$ in the fundamental equations even though in principle they are the same as our magnetogasdynamic equations (4.28) to (4.32). We list these electromagnetogasdynamic equations as follows:

$$p = R_p \rho T \tag{4.55a}$$

$$\frac{\partial \rho}{\partial t} + \nabla \cdot (\rho \vec{u}) = 0 \tag{4.55b}$$

$$\rho \frac{D\vec{u}}{Dt} = -\nabla p + \mu_e \vec{J} \times \vec{H} + (\nabla \cdot \tau) \tag{4.55c}$$

$$\rho \frac{D C_v T}{Dt} = -p(\nabla \cdot \vec{u}) + (\tau \cdot \nabla) \cdot \vec{u} + \nabla \cdot (\varkappa \nabla T) + \frac{J^2}{\sigma} \tag{4.55d}$$

$$\nabla \times \vec{H} = \vec{J} \tag{4.55e}$$

$$\nabla \times \vec{E} = -\frac{\partial \mu_e \vec{H}}{\partial t} \tag{4.55f}$$

$$\vec{J} = \sigma (\vec{E} + \mu_e \vec{u} \times \vec{H}) \tag{4.55g}$$

$$\nabla \cdot \vec{H} = 0 \tag{4.55h}$$

$$\nabla \cdot \vec{E} = 0 \tag{4.55i}$$

where the electric and the magnetic fields include both the external and the induced fields. Equations (4.55h) and (4.55i) are not independent relations [cf. equations (4.9) and (4.10)].

(d) Radiation-magnetogasdynamics. At very high temperature the radiation terms p_R and E_R in the fundamental equations may not be negligible. In this case we should include these terms in the fundamental equations. Including these radiation terms will not affect the forms of other terms in the fundamental equations. With radiation terms, we may still apply the magnetogasdynamic approximations or other simplifications which we discussed before. The fundamental equations for radiation-magnetogasdynamics are as follows:

$$p = R_p \rho T \tag{4.56a}$$

$$\frac{\partial \rho}{\partial t} + \nabla \cdot (\rho \vec{u}) = 0 \tag{4.57b}$$

$$\rho \frac{D\vec{u}}{Dt} = -\nabla (p + p_R) + \mu_e \vec{J} \times \vec{H} + (\nabla \cdot \tau) \tag{4.56c}$$

$$\rho \frac{D}{Dt}\left(C_v T + \frac{E_R}{\rho}\right) =$$
$$= -(p + p_R)(\nabla \cdot \vec{u}) + (\tau \cdot \nabla) \cdot \vec{u} + \nabla \cdot (\varkappa \nabla T) + \frac{J^2}{\sigma} + \nabla \cdot (D_R \nabla E_R) \tag{4.56d}$$

$$\frac{\partial \vec{H}}{\partial t} = \nabla \times (\vec{u} \times \vec{H}) - \nabla \times [\nu_H (\nabla \times \vec{H})] \tag{4.56e}$$

where p_R, E_R and D_R are defined in equations (2.36), (2.39) and (2.45) respectively.

Of course we may apply those simplifications discussed in § 4 to the fundamental equations of radiation-magnetogasdynamics (4.56) too.

(e) Magnetogasdynamics in gravitation field. In some problems the non-electric force $F_g{}^i$ such as the gravitational force may not be neglected. In that case we should include the force $F_g{}^i$ in the equation of motion (cf. chapter VII, § 7). Sometimes an extra condition is required to determine the gravitational potential (*2*). We shall not discuss it in this book.

References

1. Alfvén, H.: On the Existence of Electromagnetic-hydrodynamic Waves. Arkiv F. mat. astr. o. fysik, Bd. 29B, No. 2, 1942.

2. Chandrasekhar, S.: An Introduction to the Study of Stellar Structure. Dover Publications, Inc., 1957.

3. Friedrichs, K. O.: Nonlinear Wave Motion in Magneto-hydrodynamics. Los Alamos Laboratory Report, 1954.

4. Hartmann, J.: Hg-dynamics. I. Kgl. Danske Vidensk. Selskab Math.-fys. Medd. XV: 6, 1937.

5. de Hoffmann, F., and E. Teller: Magnetogasdynamic Shock. Phys. Rev., vol. 80, No. 4, Nov. 15, 1950, pp. 692–703.

6. Lundquist, S.: Studies in Magneto-hydrodynamics. Arkiv F. Fysik, Bd. 5, No. 15, 1952, pp. 297–347.

7. Pai, S. I.: Energy Equation of Magneto-gasdynamics. Phys. Rev., vol. 105, No. 5, March 1, 1957, pp. 1424–1426.

8. Pai, S. I.: Introduction to the Theory of Compressible Flow. Van Nostrand Co., Inc., Princeton, N. J., 1958.

9. Pai, S. I.: Viscous Flow Theory, I-Laminar Flow. Van Nostrand Co., Inc., Princeton, N. J., 1956.

10. Stratton, J. A.: Electromagnetic Theory. McGraw-Hill Book Company, Inc., New York, 1941.

Chapter V

Dimensional Analysis and Important Parameters

1. Introduction. In chapter IV, we discussed the fundamental equations of magnetogasdynamics and other cases of plasma dynamics. These fundamental equations are nonlinear. There is no general method of finding the solutions of these nonlinear equations. In order to bring out the essential features of the flow problem, it is desirable to find important parameters which characterize the flow problem. These parameters are very useful in the analysis of experimental results. There are two methods (*4*) to find these parameters: one is known as inspection analysis and the other, dimensional analysis. In the inspection analysis, we reduce the fundamental equations to non-dimensional form and obtain the non-dimensional parameters from the resultant equations as we did in chapter IV, § 4. In dimensional analysis we form non-dimensional parameters from the physical quantities occurring in the flow problem. Both methods give the same results. In this chapter we shall use dimensional analysis to find the important non-dimensional parameters of plasma dynamics.

In § 2, we shall discuss the physical quantities in the flow problem of a fully ionized plasma. These results may be generalized to the general plasma of n-species in a straight forward manner. Some typical values of these quantities will be given.

From the physical quantities of plasma dynamics, we may form various characteristic velocities, lengths and frequencies which have important physical significance in plasma dynamics. We shall discuss these terms in § 3.

In § 4, the important parameters of a fully ionized plasma are obtained and discussed, while in § 5, we show in what range of these parameters the fundamental equations of a fully ionized plasma may be reduced to simpler cases such as magnetogasdynamics, electromagnetogasdynamics, radiation-magnetogasdynamics and other cases.

2. Fully ionized plasma (*5*). We take the following 18 variables to describe the flow problems of a fully ionized plasma (cf. chapter II, § 9):

$$T,\ p,\ \rho,\ \vec{u},\ \vec{E},\ \vec{H},\ T_1,\ p_1,\ \rho_e \text{ and } \vec{J}$$

where subscript 1 refers to the values of the electron variables. The fundamental equations for these variables are as follows:

$$p = R_p\,\rho\,T \tag{5.1a}$$

$$p_1 = R_A\,\nu_1\,T_1 \tag{5.1b}$$

$$\frac{\partial\rho}{\partial t} + \nabla\cdot(\rho\,\vec{u}) = 0 \tag{5.1c}$$

$$\rho \frac{D\vec{u}}{Dt} = -\nabla p_t + (\nabla \cdot \tau) + \vec{F}_e + \vec{F}_g \tag{5.1d}$$

$$\rho \frac{D\overline{e_m}}{Dt} = -\nabla \cdot (\vec{u}\, p_t) + \nabla \cdot (\vec{u} \cdot \tau) + \nabla \cdot \vec{Q} \tag{5.1e}$$

$$\frac{\partial \overline{e_1}}{\partial t} + \nabla \cdot [\overline{e_1 \vec{u}_1} - (\vec{u}_1 \cdot \tau_1) - \vec{Q}_1] = \varepsilon_1 \tag{5.1f}$$

$$\nabla \times \vec{H} = \vec{J} + \frac{\partial \varepsilon \vec{E}}{\partial t} \tag{5.1g}$$

$$\nabla \times \vec{E} = -\frac{\partial \mu_e \vec{H}}{\partial t} \tag{5.1h}$$

$$\frac{\partial \rho_e}{\partial t} + \nabla \cdot \vec{J} = 0 \tag{5.1i}$$

$$\vec{J} = \sigma (\vec{E} + \mu_e \vec{u} \times \vec{H}) + \rho_e \vec{u} \tag{5.1j}$$

In equations (5.1) we neglect the energy source due to chemical reaction and replace the more general current equation (2.53) by equation (2.55), i. e., (**5.1j**). The reason for using equation (2.55) is that we introduce the electrical conductivity σ which is a rather important quantity in most of the flow problems of plasma dynamics. From equation (5.1), we have the following typical physical quantities in our problem:

(1) Length L which characterizes the dimension of our flow field. All the x^i may be expressed in terms of L. The typical length L differs greatly among different flow problems. In astrophysical problem, L may be of order of 10^6 meters while in laboratory, L may be of the order of one meter. As a result, the non-dimensional parameters based on L for these two cases may be greatly different from eacy other.

(2) Velocity U. In general we have two different velocities in our problem which are the plasma velocity u^i and the diffusion velocity $w_1{}^i$. We may consider the value U as a characteristic velocity for u^i. We may use another value w to characterize $w_1{}^i$ so that $\left(\frac{w}{U}\right)$ is one of the non-dimensional parameters in our problem. Another way is to express the diffusion velocity in terms of a diffusion coefficient $\mathfrak{D}$ which will be discussed later. The typical velocity U also differs greatly among different flow problems. The typical speed in some astrophysical problems or of a satellite may be as high as of the order of 10^4 meters per second, while in laboratory test of the flow of a conducting liquid such as mercury may be as low as of the order of 1 meter per second.

(3) Time t_0 which characterizes the interval considered. The time t in equations (5.1) may be expressed in terms of this value t_0. As we mentioned in chapter IV, § 2, we may take the time scale t_0 of the order of L/U. For instance, if L is one meter and U is 100 m/sec., t_0 will be of the order of 10^{-2} sec. However in many high frequency phenomena, t_0 may be as low as of the order of 10^{-6} sec. For magnetogasdynamic approximations, we should neglect all the phenomena of the time scale much smaller than 10^{-2} second for the case of $L = 1$ m. and $U = 100$ m/sec.

(4) Number density $\nu = \nu_1 + \nu_2$ of the plasma. In addition to ν, we have the non-dimensional parameter $c_1 = \nu_1/\nu$ which is the concentration of electrons in the plasma. For air at normal temperature and pressure, the number density is about $2.7 \cdot 10^{19}$ molecules per cubic centimeter. In fusion reaction at rather low density, the number density is of the order of 10^{15} molecules per cubic centimeter. In ordinary laboratory test, the number density which we are interested in may vary from 10^{12} to 10^{20} molecules per cubic centimeter while in astrophysical problem the number density may be much smaller than these values.

(5) Mass m of the plasma defined by equation (2.19). In addition to mass m of the plasma as a whole, we have also the non-dimensional parameter m_1/m in our problem. The mass of an electron m_1 is about $9.1 \cdot 10^{-28}$ gr. while that of a proton m_2 is about $1.67 \cdot 10^{-24}$ gr. The ratio of these two masses m_1/m_2 is about 1/1840. For a fully ionized plasma consisting of electrons and protons (i. e., hydrogen ions), the mass of the plasma as a whole is about the ½ of that of a proton, i. e., $m = \frac{1}{2} m_2$. Because of the smallness of the ratio of m_1/m_2 many approximations may be made in the treatment of the flow problem of a plasma such as we discussed in chapter III, § 4.

(6) Temperature T of the plasma. In addition to T, we have T_1/T as a non-dimensional parameter. The typical temperature T_0 in various flow problems may differ greatly. In fusion reaction, typical temperature may be as high as of the order of 10^9 °K. Ordinary gas will be ionized at a temperature above 10^4 °K. For laboratory test of the flow of a conducting liquid, the temperature may be just the room temperature which is only of the order of 300 °K.

(7) Pressure p of the plasma. In addition to p, we have p_1/p as a non-dimensional parameter. The typical pressure p_0 in various flow problem may also differ greatly. In fusion reaction, the pressure of the plasma may be as high as 10^3 atmospheres ($7.6 \cdot 10^5$ mm. Hg) while the initial pressure in a shock tube may be as low as a few mm. Hg.

(8) Boltzmann constant R_A which characterizes the gas constant of the plasma and that of its species. R_A is $1.379 \cdot 10^{-16}$ erg/deg. The gas constant of the plasma is $R_p = R_A/m$.

(9) Coefficient of viscosity μ which characterizes the viscous stress of the plasma. Since the coefficient of viscosity of a species in the plasma may not be equal to that of the plasma as a whole, we have μ_1/μ as a non-dimensional parameter (*1*). The typical value of coefficient of viscosity is different for different conducting fluid. For mercury at room temperature, μ is $1.5 \cdot 10^{-2}$ g/cm-sec. In standard atmosphere, μ is $1.8 \cdot 10^{-4}$ g/cm-sec. For ionized air at an altitude of $3.5 \cdot 10^5$ ft., μ is about $2 \cdot 10^{-4}$ g/cm-sec. For ionized argon in a shock tube μ may be as high as $1.8 \cdot 10^3$ g/cm-sec.

(10) Non-electric force F_g such as gravitational force and other force. The constant of gravitation G is $6.67 \cdot 10^{-8}$ dynes cm²/g² such that the attractive force between two masses m_1 and m_2 (in g.) at a distance r cm. is $F_g = G\, m_1\, m_2/r^2$.

(11) Specific heat at constant volume C_v which characterizes the internal energy of the plasma. We have also C_{v1}/C_v as a non-dimensional parameter. For a monatomic gas, $C_v = \frac{3}{2} R_A/m$ erg/deg-g. where m is the mass of the gas in grams.

(12) Coefficient of diffusion $\mathfrak{D}$. It is sometimes convenient to express the

diffusion velocity $w_1{}^i$ in terms of the coefficient of diffusion $\mathfrak{D}$ and the gradient of concentration. For a first approximation, the diffusion coefficient may be used to characterize the diffusion velocity. For a plasma, the diffusion coefficient may depend on the constituents of the plasma as well as the magnetic field strength (*1*). In order to give some order of magnitude of the diffusion coefficient, we mention two cases of neutral gases. (a) Air-O_2, $\mathfrak{D} = 0.17778$ cm²/sec. and (b) H_2-O_2, $\mathfrak{D} = 0.66550$ cm²/sec. both values are given at 0 °C and 760 mm. Hg pressure.

(13) Coefficient of heat conductivity $\varkappa$ which characterizes the heat conduction of the plasma $Q_c{}^i$. We have also $\varkappa_1/\varkappa$ as a non-dimensional parameter. A typical value of the coefficient of heat conductivity $\varkappa$ for air at 0 °C. is $0.295 \cdot 10^{-3}$ cal/cm-sec. °K.

(14) Radiation constant a_R which characterizes both the radiation pressure and the radiation energy. The value of a_R is $7.67 \cdot 10^{-15}$ erg. cm^{-3} °K^{-4}.

(15) Diffusion coefficient of radiation D_R which characterizes the radiation flux.

(16) Magnetic field strength H. A magnetic field strength of about $2 \cdot 10^4$ gausses saturates an iron core at room temperature. A good permanent magnet has a strength of the order of $2 \cdot 10^3$ gausses. It is reasonable to take a typical value of magnetic field strength H_0 as $1 \cdot 10^3$ gausses to estimate some of the parameters.

(17) Electric field strength E. The typical value of electric field strength is rather arbitrary. In many problems of magnetogasdynamics, we take it as zero. Another typical value is that $E_0 = 0\,(U\,B_0)$.

(18) The absolute electric charge e. Its value is $e = 4.8 \cdot 10^{-10}$ esu $= 1.6 \cdot 10^{-20}$ emu $= 1.6 \cdot 10^{-19}$ coulomb.

(19) Magnetic permeability μ_e. The value of μ_e in free space in MKS units is $4\,\pi \cdot 10^{-7}$ kg-m-(coulomb)$^{-2}$.

(20) Inductive capacity ε. The value of ε in free space in MKS units is $8.854 \cdot 10^{-12}$ (coulomb)² sec² kg^{-1} m^{-3}.

(21) Electrical conductivity σ which characterizes the Joule heat. The electrical conductivity of various conducting fluids has a large range of values. For mercury, $\sigma = 10{,}800$ mho/cm., for salt water (saturated at 25 °C.), $\sigma = 0.25$ mho/cm. and for air behind a normal shock wave of Mach number 15, $\sigma \simeq 1.0$ mho/cm.

(22) Friction coefficient α which characteristes the electrical interaction force F_{e10} defined in equation (3.19). The value of α is closely associated with σ as shown in equation (3.21).

In conclusion we may use the above 22 quantities to characterize our problem of plasma dynamics in addition to such non-dimensional quantities as c_1, m_1/m etc. discussed above. In generalizing the present results to a plasma of n-species, we need only to add chemical reaction term and mass source term to these 22 quantities and to increase those non-dimensional quantities c_1 etc. from one item to $n - 1$ items, e. g., c_r with $r = 1, 2, \ldots n - 1$ etc. Hence our present results will cover essentially all the main points of a plasma in general.

3. Some characteristic velocities, frequencies and lengths in plasma dynamics. Before we are going to find those non-dimensional parameters from the above 22 quantities which characterize plasma dynamics, we list first some well known velocities, frequencies and lengths which may be derived from these 22 quantities and which are useful in our discussions of the non-dimensional parameters later (*3, 4*).

(I) Characteristic velocities.

(1) Velocity of light $= c = \dfrac{1}{\sqrt{\varepsilon \mu_e}}$ (5.2)

This is a characteristic velocity of electromagnetic waves. In free space, we have

$$c = 3.0 \cdot 10^8 \text{ m/sec.} \tag{5.2a}$$

(2) Velocity of sound $= a = \sqrt{\gamma R_p T}$ (5.3)

This is a characteristic velocity of sound wave propagation in ordinary gas-dynamics. For monatomic gas of $m = 1.67 \,.\, 10^{-24}$ and $\gamma = \frac{5}{3}$, we have

$$a = 1.18\, T \cdot 10^2 \text{ m/sec.} \tag{5.3a}$$

If $T = 10^4$ °K., then $a = 1.18 \cdot 10^4$ m/sec. For air at ordinary temperature (not ionized), $a = 340$ m/sec.

(3) Velocity of Alfvén's wave $= V_H = H \sqrt{\dfrac{\mu_e}{\rho}}$ (5.4)

This is a characteristic velocity of waves in magnetogasdynamics. For example, if we take a typical value of magnetic induction $B_0 = 10^3$ gausses $= 10^{-1}$ weber/m^2, equation (5.4) becomes

$$V_H = 89.5 \sqrt{\frac{1}{\rho}} \text{ m/sec.} \tag{5.4a}$$

where ρ is in kg/m^3. For mercury, $\rho = 13.6 \cdot 10^3$ kg/m^3, equation (5.4a) gives $V_H = 0.77$ m/sec., while for ionized air at about 100 km. altitude $\rho = 2 \cdot 10^{-6}$ kg/cm^3, equation (5.4a) gives $V_H = 6.35 \cdot 10^4$ m/sec.

We shall show in chapter X that these velocities are the basic characteristic velocities of waves in a plasma. The general velocity of wave propagation in a plasma is obtained from the interaction of these characteristic velocities and other physical quantities of the plasma.

(II) Characteristic frequencies.

(1) Plasma frequency $= \omega_p = e \sqrt{\dfrac{\nu}{m \varepsilon}}$ (5.5)

This is the frequency of electrostatic plasma oscillation. In other words, it is the frequency of oscillation due to the electric field alone. The existence of plasma oscillations was first predicted and observed by Tonks and Langmuir (*9*). Now if we take $\nu = 10^{16}$/cc. $= 10^{22}$/m^3, equation (5.5) becomes

$$\omega_p = 5.38 \frac{1}{\sqrt{m}} 10^{-3} \text{ sec}^{-1} \tag{5.5a}$$

where m is the mass of a particle in kg. For an electron $m_e = 9.1 \cdot 10^{-31}$ kg., we have from (5.5a)

$$\omega_{pe} = 5.62 \cdot 10^{12} \text{ sec}^{-1}$$

For an ionized hydrogen atom, $m_i = 1.67 \cdot 10^{-27}$ kg., equation (5.5a) gives

$$\omega_{pi} = 1.3 \cdot 10^{11}\ \text{sec}^{-1}$$

(2) Cyclotron frequency $= \omega_c = \dfrac{e\,\mu_e H}{m} = \dfrac{e B}{m}$ (5.6)

This is equal to the angular frequency with which particles of mass m and charge e gyrate in a cyclotron. Of course, we may define the cyclotron frequencies for ions or electron by using their proper values of mass and charge. It is also known as Larmor frequency. If we take a typical value of magnetic induction $B = 10^3$ gausses $= 10^{-1}$ weber/m^2, equation (5.6) becomes

$$\omega_c = \frac{1}{m} 1.6 \cdot 10^{-20}\ \text{sec}^{-1} \tag{5.6a}$$

If $m = m_e$, the electron cyclotron frequency is $\omega_{ce} = 1.76 \cdot 10^{10}$ sec^{-1}. If $m = m_i$, the ion frequency cyclotron is $\omega_{ci} = 9.6 \cdot 10^6$ sec^{-1}.

(3) Electrical decay frequency $= \omega_e = \dfrac{J}{D}$ (5.7)

If we take Ohm's law of the form $\sigma \vec{E} = \vec{J}$, equation (5.7) becomes

$$\omega_e = \frac{\sigma}{\varepsilon} = c^2 \sigma \mu_e \tag{5.8}$$

If we take $\sigma = 1$ mho/cm. $= 100$ mho/m., we have $\omega_e = 3.77 \cdot 10^4$ sec^{-1}. Equation (5.8) is well-known in the electromagnetic theory. It prescribes the rate of electrical energy converted into Joule heat.

(4) Friction frequency $= \omega_f = \dfrac{\alpha}{m \nu}$ (5.9)

This is the frequency of the oscillation if the friction force F_{e10} is the only external force. When we use the approximate formula for α, i. e.,

$$\alpha = \frac{\nu^2 e^2}{\sigma} \tag{5.10}$$

equation (5.9) gives $\omega_f = 1.53 \cdot 10^3$ sec^{-1}, where we take $\nu = 10^{16}$/cc., $\sigma = 1$ mho/cm. and $m = m_i = 1.67 \cdot 10^{-27}$ kg.

(III) Characteristic lengths.

(1) Debye length $= L_D = \sqrt{\dfrac{R_A T \varepsilon}{\nu e^2}}$ (5.11)

It is obtained from the condition that the electrical potential energy is equal to the thermal energy. In the analysis of an electrolyte, Debye length was originally obtained as the distance that the electric field of a point charge is shielded by a local increase of concentration of charges of opposite sign. In a plasma, it is a measure of the distance over which the excess electrical charge may be different appreciably from zero.

Equation (5.11) may be written as

$$L_D = 4.75 \sqrt{\frac{T}{\nu}}\, 10^3 \text{ meters} \tag{5.11a}$$

where T is in °K. and ν is in molecules per cubic meter. If we take $T = 10^4$ °K., $\nu = 10^{22}/\text{m}^3$, equation (5.11a) gives

$$L_D = 4.75 \cdot 10^{-6} \text{ m.} = 4.75 \cdot 10^{-3} \text{ mm.} \tag{5.11b}$$

We see that the Debye length is very small and thus the plasma has its tendency toward electrically neutrality, i. e., ρ_e tends to be zero. In the analysis of magnetogasdynamics, $\rho_e = 0$ is a good approximation.

(2) Mean free path $= L_f = 1.255 \sqrt{\gamma}\, \frac{\mu}{a\rho}$ (5.12)

It is a measure of distance travelled between collisions of neutral particles. For a plasma or charge particles, the mean free path depends also on the electric and magnetic fields (*1, 3*). When the typical length of the flow field is much larger than the mean free path, the continuum theory holds true. If the typical length of the flow field is of the order of the mean free path or smaller, we should use the microscopic treatment.

For air $\gamma = 1.4$. In standard atmosphere, we may take $\mu = 1.8 \cdot 10^{-4}$ g/cm-sec., $\rho = 1.23 \cdot 10^{-3}$ g/cm³, and $a = 340$ m/sec.; then we have

$$L_f = 0.64 \cdot 10^{-5} \text{ cm.}$$

Ordinarily the mean free path is very small. As a result, the continuum theory of gasdynamics is very good.

At an altitude of 100 km., we have $\mu = 2 \cdot 10^{-4}$ g/cm-sec., $\rho = 2 \cdot 10^{-9}$ g/cc. and $a = 400$ m/sec., we have

$$L_f = 3.7 \text{ cm.}$$

The mean free path increases with the altitude. At very high altitude, we should watch the ratio of L/L_f before we apply the continuum theory of gasdynamics.

(3) Larmor radius $= L_L =$ radius of gyration $= \frac{a}{\omega_c} = \sqrt{\gamma R_p T}\, \frac{m}{eB}$ (5.13)

In equation (5.13), we use the sound velocity "a" to represent the particle velocity of the plasma. It is a measure of the radius of the helical path of a charged particle of mass m and charge e in a magnetic field H. From our previous example, we have $a = 1.18 \cdot 10^4$ m/sec., $\omega_c = 9.6 \cdot 10^6$ sec⁻¹, and then the corresponding Larmour radius is

$$L_L = 1.23 \cdot 10^{-3} \text{ m.} = 1.23 \text{ mm.}$$

We have shown some typical values of these characteristic velocities, frequencies and lengths. We shall discuss their physical significance further in the following chapters.

4. Important parameters of plasma dynamics. We are going to obtain the non-dimensional parameters which characterize the flow of a fully ionized plasma by dimensional analysis from the 22 physical quantities given in § 2. In our problem, there are five independent fundamental units, i. e., mass, time, length, temperature and electrical charge. By the theorem of dimensional analysis (*5*), we may form 17 non-dimensional parameters to characterize our problem from these 22 quantities. The choice of these 17 non-dimensional parameters is arbitrary; but there are only 17 independent non-dimensional parameters. We shall write down these non-dimensional parameters according to their special physical significance including those forms well-known in gasdynamics and electrodynamics. It is convenient to divide these parameters into two groups: (A) parameters of ordinary gasdynamics and (B) new parameters of plasma dynamics. We take the length L, mass m, time t_0, temperature T and electrical charge e as the five fundamental units of our system. We derive non-dimensional parameters of all other quantities listed in 2 in terms of these fundamental units. We shall call these non-dimensional parameters as Π_n, because of the well-known Π-theorem in dimensional analysis (*5*). We shall also identify these Π_n with those well-known parameters of gasdynamics and electrodynamics.

(A) Parameters of ordinary gasdynamics.

(1) Velocity U. The non-dimensional parameter is

$$\Pi_1 = R_t = \frac{U t_0}{L} \tag{5.14}$$

This parameter characterizes the time scale of the problem with respect to the flow velocity. In ordinary gasdynamics or magnetogasdynamics, we consider only the cases where R_t is of the order of unity or larger. But in high frequency communication problems, R_t may be very small.

(2) Pressure p. The non-dimensional parameter is

$$\Pi_2 = \frac{p}{\frac{\rho U^2}{2}} \tag{5.15}$$

This is the ratio of the pressure of the plasma to the dynamic pressure.

(3) Ratio of specific heats γ [cf. equation (4.33)]

$$\Pi_3 = \gamma = \frac{C_p}{C_v} = 1 + \frac{R_p}{C_v} = 1 + \frac{R_A}{m C_v} \tag{5.16}$$

For monatomic gas $m C_v = \frac{3}{2} R_A$, hence $\gamma = \frac{5}{3}$. For diatomic gas at ordinary temperature $m C_v = \frac{5}{2} R_A$, hence $\gamma = \frac{7}{5}$.

(4) Mach number M [cf. equation (4.34)]

$$\Pi_4 = M = \frac{U}{\sqrt{\gamma R_p T}} \tag{5.17}$$

(5) Reynolds number R_e [cf. equation (4.35)]

$$\Pi_5 = R_e = \frac{\rho U L}{\mu} \tag{5.18}$$

(6) Prandtl number P_r [cf. equation (4.36)]

$$\Pi_6 = P_r = \frac{\mu C_p}{\varkappa} \tag{5.19}$$

We have already discussed the ratio of specific heats γ, Mach number M, Reynolds number R_e and Prandtl number P_r in chapter IV, § 4. Hence we will not repeat them here.

(7) Froude number F_r

$$\Pi_7 = \frac{\rho U^2}{F_g L} \tag{5.20}$$

If $F_g = \rho g$, we have

$$\Pi_7 = F_r = \frac{U^2}{g L} \tag{5.20a}$$

Froude number is a measure of the effect of the gravitational force on the flow. It is important when there is a free surface.

(8) Schmidt number S_c

$$\Pi_8 = S_c = \frac{\mu}{\rho \mathfrak{D}} \tag{5.21}$$

Schmidt number is a measure of the diffusion effect.

(B) New parameters of plasma dynamics.

(9) Relativity parameter R_c [cf. equation (4.14)]

$$\Pi_9 = R_c = \frac{U^2}{c^2} = U^2 \varepsilon \mu_e \tag{5.22}$$

This parameter is usually a very small number for the flow problem. For instance, if we take such a high flow velocity as $U = 10^4$ m/sec., R_c is only $1.11 \cdot 10^{-9}$.

(10) Electric decay parameter

$$\Pi_{10} = t_0 \omega_e = t_0 \frac{\sigma}{\varepsilon} = t_0 \frac{c^2}{\nu_H} \tag{5.23}$$

If we take typical values $t_0 = 10^{-2}$ sec. and $\omega_e = 3.77 \cdot 10^4 \text{ sec}^{-1}$, we have $\Pi_{10} = 3.77 \cdot 10^2$.

(11) Plasma frequency parameter

$$\Pi_{11} = t_0 \omega_p = t_0 e \sqrt{\frac{\nu}{m \varepsilon}} \tag{5.24}$$

As an example, we take $t_0 = 10^{-2}$ sec., $\omega_{pi} = 1.3 \cdot 10^{11} \text{ sec}^{-1}$, we have $\Pi_{11} = 1.3 \cdot 10^9$. For ordinary flow problem, this parameter is always a very large number.

(12) Frictional frequency parameter

$$\Pi_{12} = t_0 \omega_f = t_0 \frac{\alpha}{m \nu} \tag{5.25}$$

As an example, we take $t_0 = 10^{-2}$ sec., $\omega_f = 1.53 \cdot 10^3$, we have $\Pi_{12} = 15.3$.

The last three parameters Π_{10}, Π_{11}, and Π_{12} show the time scale of the flow field with respect to the frequencies ω_e, ω_p and ω_f respectively.

(13) Electrical field parameter R_E [cf. equation (4.12)]

$$\Pi_{13} = R_E = \frac{E}{U\,\mu_e\,H} \tag{5.26}$$

As we have shown in chapter IV, when the electrical conductivity is very large, we expect that R_E is of the order of unity. For many problems of flow of fluid of low electrical conductivity, we may assume that $E = 0$ then $R_E = 0$ (cf. chapter VI).

(14) Magnetic pressure number R_H [cf. equations (4.37)]

$$\Pi_{14} = R_H = \frac{\mu_e H^2}{\rho\,U^2} = \frac{V_H{}^2}{U^2} \tag{5.27}$$

From our previous numerical example, for mercury we have $V_H = 0.77$ m/sec. If we take a typical flow velocity of 1 m/sec., we have $R_H = 0.593$. For ionized air at about 100 km. altitude, we have $V_H = 6.35 \cdot 10^4$ m/sec. If in this case we take a typical speed of a space vehicle, $U = 10^4$ m/sec. we have $R_H = 40$. This parameter is large for flow over a space vehicle.

(15) Magnetic Reynolds number R_σ [cf. equation (4.38)]

$$\Pi_{15} = R_\sigma = U\,L\,\sigma\,\mu_e = \frac{U\,L}{\nu_H} \tag{5.28}$$

If we take $\sigma = 100$ mho/m., equation (5.28) in MKS units becomes

$$R_\sigma = 4\,\pi \cdot 10^{-5}\,U\,L \tag{5.28a}$$

If we take $U = 10^4$ m/sec., $L = 1$ m., $R_\sigma = 1.264$. For magnetoaerodynamics, the value of R_σ is usually very small.

(16) Radiation pressure parameter (cf. chapter XIII, § 7)

$$\Pi_{16} = \frac{p_R}{p} = \frac{\frac{1}{3} a_R\,T^4}{p} \tag{5.29}$$

This is the ratio of the radiation pressure to the gas pressure of the plasma. For $T = 10^4$ °K., $p_R = 2.5 \cdot 10^{-5}$ atm., and for $T = 10^5$ °K., $p_R = 0.25$ atm. For moderate high pressure temperature, such as $T = 10^4$ °K., the radiation pressure parameter is usually a very small number. Only at very high temperature and low gas pressure, this radiation pressure parameter is not small and then the effect of radiation pressure on the flow problem will not be negligible.

(17) Radiation flux parameter

$$\Pi_{17} = \frac{D_R\,a_R\,T^4}{\varkappa\,T} = \frac{D_R\,a_R}{\varkappa}\,T^3 \tag{5.30}$$

It is a measure of the relative importance between heat transfer due to radiation and that due heat conduction.

From the parameters Π_{16} and Π_{17}, we see that the radiation phenomena are important at very high temperature because these radiation parameters are proportional to T^3 for a given density of the plasma.

5. Important parameters for special cases. Even though in the most general cases of plasma dynamics, there are 17 important parameters as shown in the last section, for each special case only a few of these 17 parameters are important. We are going to discuss a few special cases in this section.

(a) Magnetogasdynamics. As we have shown in chapter IV, § 2, for magnetogasdynamics, the 17 parameters may be divided into the following three groups:

(i) Those having definite values: For magnetogasdynamics, we have

$$R_t \simeq 1 \text{ (or } R_t > 1), \quad R_E \cong 1 \text{ (or } R_E < 1), \quad R_c \ll 1$$

This means that in the problems of magnetogasdynamics, we consider only those problems in which the time scale t_0 is of the order of U/L or larger and the electric field E is of the order of UB or smaller and the flow velocity is much smaller than the light velocity.

(ii) Important parameters which have considerable influence on the flow problems. For magnetogasdynamics, they are seven:

$$\Pi_2, \ \gamma, \ M, \ R_e, \ P_r, \ R_H, \text{ and } R_\sigma$$

In a general investigation of magnetodynamics, we should consider the effects of these parameters on the flow problem. In different range of these parameters, we have to use different analysis of the flow problem. For instance, if $R_e = R_\sigma = P_r = \infty$, we have the case of ideal plasma. We shall discuss these problems in the following chapters.

(iii) Unimportant parameters which may be neglected in the treatment of ordinary magnetogasdynamics and which are the remaining seven:

$$F_r, \ S_c, \ \Pi_{10}, \ \Pi_{11}, \ \Pi_{12}, \ \Pi_{16}, \text{ and } \Pi_{17}$$

(b) Magnetohydrodynamics. It is the same as magnetogasdynamics except that in the present case we take $M \simeq 0$. It should be noticed that the parameters γ and P_r are still important parameters when we study the temperature distribution for the case $M \simeq 0$. If we are interested only in the velocity distribution, these two parameters may be neglected.

(c) Electromagnetogasdynamics. It is the same as magnetogasdynamics except that in the present case we may replace the condition $R_E \simeq 1$ by $R_E \leqq c/U$. As a result the energy in the electric field may be of the same order of magnitude as that in the magnetic field but the fundamental equations remain the same as those of magnetogasdynamics.

(d) Radiation magnetogasdynamics. It is the same as magnetogasdynamics except that the parameters of radiation Π_{16} and Π_{17} are no longer negligibly small. We should retain the radiation terms in the fundamental equations [cf. equation (4.56)]. This is the case when the average temperature of the flow is very high, i. e., $T > 10^5$ °K.

(e) Magnetogasdynamics in gravitational field. It is the same as magnetogasdynamics except that the Froude number should be considered as an important parameter.

(f) Wave motions in a plasma. Here the parameters may be divided into the following groups:

(i) $R_c \ll 1$.

In ordinary plasma dynamics, this condition is always satisfied so that the relativistic effects may be neglected.

(ii) Important parameters are R_t, R_E, Π_2, γ, M, R_e, R_H, P_r, R_σ, Π_{10}, Π_{11}, and Π_{12}.

In a general investigation of wave motion, the time scale t_0 may be very small such that in the very high frequency range. As a result, the frequency parameters Π_{10}, Π_{11} and Π_{12} may become important.

(iii) Unimportant parameters are F_r, S_c, Π_{16}, and Π_{17}.

It should be noticed that in the above cases (a) to (f), we may have some further simplified cases. For instance, if we consider the wave motions in an ideal plasma, the effects of the parameters R_e, P_r, R_σ, Π_{10}, and Π_{12} are negligible. Another point which should be mentioned is that in some problem a combination of some of the 17 parameters may be regarded as a new important parameter. One example is that

$$\text{Hartmann number} = R_h = \sqrt{R_e R_H R_\sigma} \tag{5.31}$$

The following gives some other important example of such combinations:

(g) Continuum theory vs kinetic theory (*1*, *4*). It is well-known that for ordinary gasdynamics, the continum theory of Navier-Stokes equations holds only when the mean free path is much smaller than the dimension of the flow field. When the mean free path is not small in comparison with the dimension of the flow field, the discrete character of the particles of the fluid in the flow should be taken into consideration. The non-dimensional parameter for this case is obviously

$$R_L = \frac{L_f}{L} = 1.255 \sqrt{\gamma}\, \frac{M}{R_e} \tag{5.32}$$

The parameter R_L known as Knudsen number depends on the three parameters, γ, M and R_e. Since the variation of the value γ is small, usually we consider the Knudsen number R_L as a function of Mach number M and Reynolds number R_e, i. e., it increases as M increases and decreases as R_e increases (*10*). From our previous numerical example (5.12) we see that in standard atmosphere the Knudsen number is usually very small, e. g., if we take $L = 1$ cm., $R_L = 0.164 \cdot 10^{-5}$. On the other hand, at high altitude, we should be careful about the application of the continuum theory in the analysis of flow problems. The same thing is true for the case of low density plasma.

Another important parameter which has a similar effect as R_L is the ratio of Larmor radius to the dimension of the flow field, i. e.,

$$R_{La} = \frac{L_L}{L} = \frac{R_t}{M \Pi_{11}} \sqrt{\frac{1}{R_H R_c}} \tag{5.33}$$

If R_{La} is negligibly small, we may consider the plasma as a continuum medium under the influence of a magnetic field. If R_{La} is large, we should consider the discrete effect of charged particles under the influence of a magnetic field (cf. chapter XIV). There are cases where the ion Larmor radius L_L is smaller than the mean free path L_f (cf. § 3). The collisionless shock may occur under such a condition.

References

1. Burgers, J. M.: Applications of Transfer Equations to the Calculation of Diffusion. Heat Conduction, Viscosity and Electric Conductivity. I and II. Tech. note 124 a and b, Inst. for Fluid Dynamics and Appl. Math. Univ. of Md., May 1958.
2. Chang, C. C., and J. T. Yen: Rayleigh's Problem in Magnetohydrodynamics. The Phys. of Fluids, vol. 2, no. 4, pp. 393–403, July–Aug. 1959.
3. Grad, H., and M. H. Rose: Dimensional Considerations. Notes of Magnetohydrodynamics, No. II. Inst. of Math. Sci., New York Univ., 1956.
4. Kantrowitz, A. R., and H. E. Petschek: An introductory discussion of magnetohydrodynamics. A Symposium of magnetohydrodynamics. Ed. by R. K. M. Landshoff, Stanford Univ. Press, pp. 3–15, 1957.
5. Pai, S. I.: Viscous Flow Theory I. Laminar Flow. D. Van Nostrand Co., Inc., Princeton, N. J., 1956.
6. Pai, S. I.: Plasmadynamics from Gasdynamic Point of View. Proc. of First Intern. Symp. of rarefied gasdynamics. Pergamon Press, 1959.
7. Pai, S. I., and A. Speth: The Wave Motions of Small Amplitude in Radiation-electromagneto-gasdynamics. Proc. of 6th Midwestern Conf. of Fluid Rech. Univ. of Texas, pp. 446–466, Sept. 1959.
8. Resler, E. L. Jr., and W. R. Sears: The Prospects for Magneto-aerodynamics. Jour. Aero. Sci., vol. 25, No. 4, pp. 235–245, April 1958.
9. Tonks, L., and I. Langmuir: Oscillations in Ionized Gases. Phys. Rev., vol. 33, pp. 195–210, Feb. 1929.
10. Tsien, H. S.: Superaerodynamics. Jour. Aero. Sci., vol. 13, No. 12, pp. 633–664, Dec. 1946.

Chapter VI

Magnetohydrodynamics

1. Introduction. In this chapter we are going to discuss some problems of magnetohydrodynamics whose fundamental equations are equations (4.47). Since equations (4.47) are much more complicated than ordinary Navier-Stokes equations, an extensive investigation of equation (4.47) is beyond the scope of this short book. What we are going to discuss is to give a few simple flow problems which show the essential difference between magnetohydrodynamics and ordinary hydrodynamics. In § 2, we discuss the flow between parallel plates, both the Poiseuille and Couette flows will be treated. In § 3, the flow through a pipe will be discussed. In § 4, the flow near a stagnation point will be considered.

Another interesting problem is the flow of a conducting fluid past a magnetized body. We shall consider an example of an inviscid and incompressible fluid over a magnetized sphere in § 5.

One of the most interesting class of problems of viscous flow in ordinary hydrodynamics is the boundary layer flow. When the Reynolds number is very large, the viscous effect will be confined near the boundary. In this way we may simplify the Navier-Stokes equations by boundary layer approximations so that viscous effects of many important practical problems may be evaluated. It will be interesting to see under what conditions the magnetohydrodynamic equations may be simplified according to the boundary layer approximations. We shall discuss the boundary layer flow of fluid of very small electrical conductivity in § 6 and that of large or infinite electrical conductivity in § 7.

In § 8, we discuss a simple unsteady flow of magnetohydrodynamics, i. e., the Rayleigh problem of magnetohydrodynamics.

Our main interest is the velocity and the magnetic field distributions and their interactions which are based on equations (4.47).

In § 9, we shall give a brief discussion of temperature distribution in magnetohydrodynamics based on the thermal equation (4.48) or (4.49).

2. Laminar flow between parallel plates (*2, 3*). Let us consider the two-dimensional steady laminar flow of an incompressible and electrically conducting fluid of constant viscosity and constant electrical conductivity between two parallel straight insulated walls. The velocity of the flow is parallel to the walls which are in the direction of the x-axis. There is an external magnetic field of constant strength H_0 in the y-direction. All the variables are independent of the z-coordinate and the time t. The conditions of this problem are expressible as follows:

$$u = U\, u^*(y^*), \quad v = 0, \quad w = 0$$

$$H_x = H_0\, H_x^*(y^*), \quad H_y = H_0, \quad H_z = 0 \tag{6.1}$$

$$p = \rho\, U^2\, p^*\,(x^*, y^*), \quad \frac{\partial(\,)}{\partial t} = 0, \quad \frac{\partial(\,)}{\partial z} = 0$$

where $x = L\,x^*$, $y = L\,y^*$. L and U are the characteristic length and velocity of the problem respectively. We are going to find the non-dimensional variables $u^*\,(x^*)$, $H_x^*\,(x^*)$ and $p^*\,(x^*, y^*)$ for given boundary conditions.

Substituting equation (6.1) into equation (4.47) we have the differential equations for u^*, H_x^* and p^* as follows:

$$\begin{aligned} \frac{1}{R_e}\cdot\frac{\mathrm{d}^2 u^*}{\mathrm{d}\,y^{*2}} + R_H\frac{\mathrm{d}\,H_x^*}{\mathrm{d}\,y^*} &= +\frac{\partial\,p^*}{\partial\,x^*} \\ R_H\,H_x^*\frac{\mathrm{d}\,H_x^*}{\mathrm{d}\,y^*} &= -\frac{\partial\,p^*}{\partial\,y^*} \\ \frac{\mathrm{d}\,u^*}{\mathrm{d}\,y^*} + \frac{1}{R_\sigma}\cdot\frac{\mathrm{d}^2\,H_x^*}{\mathrm{d}\,y^{*2}} &= 0 \end{aligned} \tag{6.2}$$

where $R_e = \rho\,U\,L/\mu$, $R_H = \mu_e\,H_0^2/(\rho\,U^2)$ and $R_\sigma = \mu_e\,\sigma\,U\,L$.

From equation (6.2), if we eliminate p^* and H_x^* we have

$$\frac{\mathrm{d}^3\,u^*}{\mathrm{d}\,y^{*3}} - R_h^2\frac{\mathrm{d}\,u^*}{\mathrm{d}\,y^*} = 0 \tag{6.3}$$

where $R_h^2 = R_e\,R_H\,R_\sigma$. R_h is the Hartmann number defined in equation (4.43). In the present problem, only the magnetic force and the viscous force are the important forces; hence Hartmann number R_h is the important parameter.

The solution of equation (6.3) is

$$u^* = \frac{A}{R_h}\cosh R_h\,y^* + \frac{B}{R_h}\sinh R_h\,y^* + C \tag{6.4}$$

where A, B and C are arbitrary constants to be determined by the boundary conditions.

(a) Plane Poiseuille flow (7). If we consider the two plates situated at $y^* = \pm 1$ and at rest and we choose the characteristic velocity as the velocity at the center between the two plates, the boundary conditions are

$$y^* = \pm 1,\;\; u^* = 0; \quad y^* = 0,\;\; u^* = 1 \tag{6.5}$$

Equation (6.4) under the conditions of (6.5) becomes

$$u^* = \frac{\cosh R_h - \cosh R_h\,y^*}{\cosh R_h - 1} \tag{6.6}$$

As R_h tends to zero, we have the velocity distribution of Poiseuille flow in ordinary hydrodynamics, i. e.,

$$u^* = 1 - y^{*2} \tag{6.7}$$

When R_h is very large, i. e., $R_h \to \infty$, we have

$$\lim_{R_h \to \infty} \frac{\cosh R_h\,y^* - 1}{\cosh R_h - 1} = 0 \tag{6.8}$$

except $y^* \to \pm 1$. Hence when $R_h \to \infty$

$$u^* = 1$$

except $y^* = \pm 1$. For very large value of R_h, the velocity distribution is almost constant in the central portion between the two plates and then drops very rapidly to zero near the plates. The reason is due to the fact that the magnetic field increases the total shearing stress of the flow field. Since the shearing stress due to the magnetic field is proportional to the magnetic field H_x which will be zero on the wall, near the wall the total shearing stress will be largely produced by the viscous force which results in a very large velocity gradient

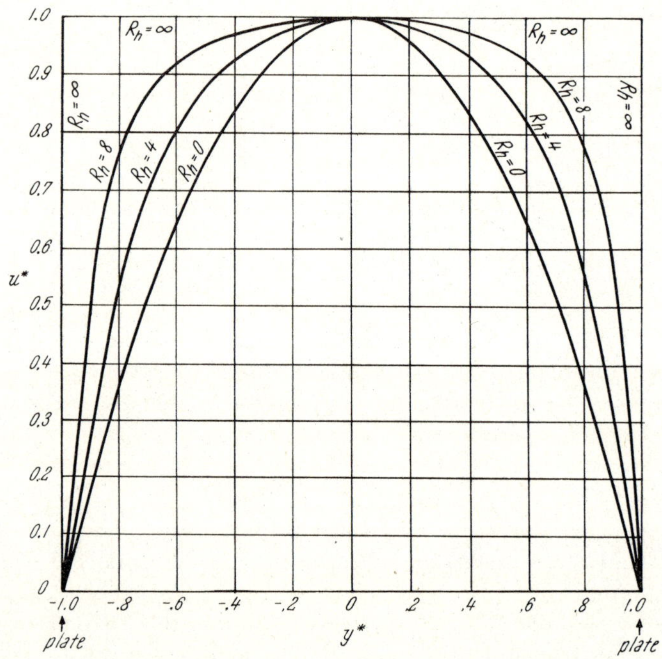

Fig. 6.1. Velocity Distributions of Plane Poiseuille Flow in Magnetohydrodynamics

near the wall. This effect of magnetic stress is qualitatively similar to that of turbulent stress which produces uniform flow in the central portion between two plates and large velocity gradient near the walls.

(b) Plane Couette flow. If we consider that the two plates are situated at $y^* = \pm 1$ and that the relative velocity between the two plates is $2\,U$ and that there is no pressure gradient in the flow field, the boundary conditions are

$$\begin{gathered} y^* = +1, \quad u^* = 1; \quad y^* = -1, \quad u^* = -1 \\ y^* = 0, \quad u^* = 0 \end{gathered} \tag{6.9}$$

Equation (6.4) under the condition of (6.9) becomes

$$u^* = \frac{\sinh R_h y^*}{\sinh R_h} \tag{6.10}$$

As R_h tends to be zero, we have the velocity distribution of Couette flow in ordinary hydrodynamics, i. e.,

$$u^* = y^* \tag{6.11}$$

When R_h is very large, i. e., $R_h \to \infty$, we have

$$\lim_{R_h \to \infty} \frac{\sinh R_h y^*}{\sinh R_h} = 0 \tag{6.12}$$

except $y^* = \pm 1$. Hence when $R_h \to \infty$

$$u^* = 0$$

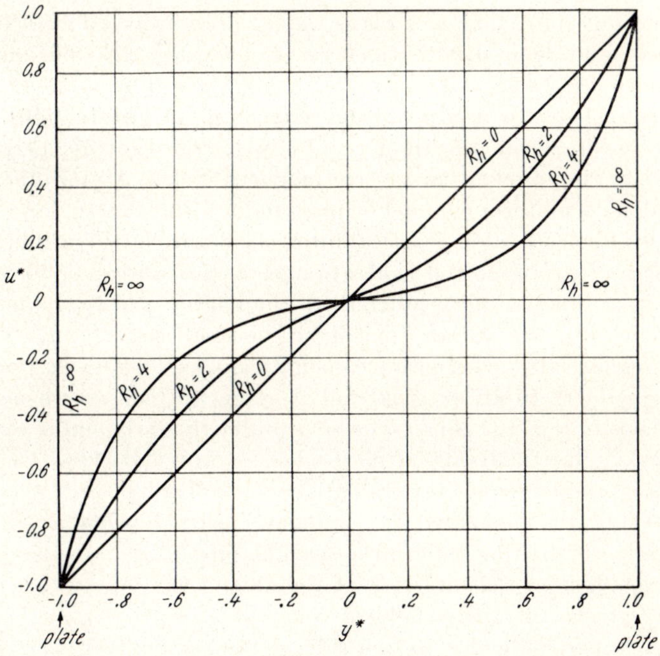

Fig. 6.2. Velocity Distributions of Plane Couette Flow in Magnetohydrodynamics

except $y^* = \pm 1$ where $u^* = \pm 1$ respectively. Here again we have uniform flow between the two plates and a very large velocity gradient near the walls.

The velocity distributions for various R_h for equations (6.6) and (6.10) are shown in figs. 6.1 and 6.2 respectively.

After the expression u^* is obtained, the values of H_x^* and p^* may be obtained by simple quadrature. The boundary conditions for H_x^* are at $y^* = \pm 1$, $H_x^* = 0$, because we consider that the plates are insulated and that originally $H_x^* = 0$ on the plates. Hence

$$H_x^* = R_\sigma \left[\frac{K_0}{2} (y^* + 1) - \int_{-1}^{y} u^* \mathrm{d} y^* \right] \tag{6.13}$$

where $K_0 = \int_{-1}^{1} u^* \, \mathrm{d}\, y^* = 0$ for plane Poiseuille flow

$$K_0 = \int_{-1}^{1} u^* \, \mathrm{d}\, y^* = \text{constant for plane Couette flow.}$$

The expression for the pressure is

$$p^* = -R_H \frac{H_x{}^{*2}}{2} + A_p \, x^* + B_p \tag{6.14}$$

where A_p and B_p are constants of integration to be determined by the boundary conditions. B_p is the pressure at $x^* = 0$ and $y^* = \pm 1$. A_p is the x-wise pressure gradient for a given problem, e. g., for plane Poiseuille flow A_p is a constant while for plane Couette flow A_p is zero.

It should be noticed that if the external magnetic field is in the direction of the plates instead of perpendicular to the plates, the velocity distributions of both plane Poiseuille flow and plane Couette flow will not be affected by the magnetic field and their distributions are exactly the distribution of ordinary hydrodynamics.

3. Laminar flow in a pipe under external magnetic field. In § 2, we discuss the case in which the fluid is infinite in the direction perpendicular to the flow and to the imposed transverse magnetic field. A more realistic picture is the flow along a uniform channel or pipe under a constant transverse magnetic field which is of considerable practical interest because of the utility of induction flow meters which rely on the generation of a measurable potential difference in the fluid in a direction perpendicular to the motion and to the magnetic field.

In reference (*8*), the author considered an ideal case of the steady laminar flow of an electrically conducting incompressible fluid in a non-conducting circular pipe subject to an external radial magnetic field which is independent of the axial position of the pipe. The solution of this problem is exactly similar to Hartmann's solution discussed in the last section. However in this case, it is difficult to produce this ideal radial magnetic field.

Shercliff (*11*) investigated the laminar steady flow of an incompressible and electrically conducting fluid on a straight pipe of rectangular cross-section. The flow velocity is in the direction of the axis of the pipe which is the x-axis. There is an external magnetic field of constant strength H_0 in the direction of y-axis which is perpendicular to one side of the rectangular section. The conditions of this problem are as follows:

$$\begin{gathered} u = u\,(y, z), \quad v = 0, \quad w = 0 \\ H_x = H_x\,(y, z), \quad H_y = H_0, \quad H_z = 0 \\ p = p\,(x, y, z), \quad \frac{\partial ()}{\partial t} = 0 \end{gathered} \tag{6.15}$$

From equation (4.47) and (6.15), we see that the pressure may be expressed in the following form:

$$p = -k\,\mu\,x + p'\,(y, z) \tag{6.16}$$

where k is a constant and $p'\,(y, z)$ is a function of y and z only which depends on $H_x\,(y, z)$. The pressure gradient along the x-axis is a constant. With the help of equation (6.16), we have to consider only u and H_x. From equations (4.47) and (6.15) we have the following equations for u and H_x:

$$\mu \nabla_1^2 u + \mu_e H_0 \frac{\partial H_x}{\partial y} + k \mu = 0$$
$$\nu_H \nabla_1^2 H_x + H_0 \frac{\partial u}{\partial y} = 0 \tag{6.17}$$

where $\nabla_1^2 = \partial^2/(\partial y^2) + \partial^2/(\partial z^2)$.

Following SHERCLIFF (*12*), we introduce the following variables instead of u and H_x:

$$V = u + (\nu_H)^{\frac{1}{2}} H_x$$
$$W = u - (\nu_H)^{\frac{1}{2}} H_x \tag{6.18}$$

From equation (6.17), we have the equations for V and W as follows:

$$\nabla_1^2 V + \frac{R_h}{L} \cdot \frac{\partial V}{\partial y} + k = 0 \tag{6.19}$$

$$\nabla_1^2 W - \frac{R_h}{L} \cdot \frac{\partial W}{\partial y} + k = 0 \tag{6.20}$$

where R_h is the Hartmann number defined in equation (4.43).

Now we assume that the boundaries of the pipe are on $y = \pm L$ and $z = \pm L_1$. Because of the symmetry of the boundary, changing the sign of y in equations (6.19) and (6.20) indicates that

$$V(y, z) = W(-y, z) \tag{6.21}$$

We need to consider V only.

We may assume that the constant k is given. Over the range $|z| < L_1$, we may express k in terms of a Fourier series as follows:

$$k = \frac{4k}{\pi} \sum \frac{(-1)^n}{2n+1} \cos \frac{(2n+1)\pi z}{2L_1} \tag{6.22}$$

this and all subsequent summations, being taken over all positive integral values of n, starting with zero. The solution of equation (6.19) is then

$$V = \frac{16 k L_1^2}{\pi^3} \sum \frac{(-1)^n}{(2n+1)^3} \left[1 + \frac{(\exp m_1 y) \sinh m_2 L - (\exp m_2 y) \sinh m_1 L}{\sinh (m_1 - m_2) L}\right] \cos \frac{(2n+1)\pi z}{2L_1} \tag{6.23}$$

where m_1 and m_2 are the roots of the quadratic

$$m^2 + \frac{R_h}{L} m - (2n+1)^2 \frac{\pi^2}{4 L_1^2} = 0 \tag{6.24}$$

The function W may be deduced from equation (6.21). After V and W are known, we may calculate u and H_x from equation (6.18). The pressure $p'(y, z)$ may be calculated from $H_x(y, z)$.

The interesting result of equation (6.23) is that for large value of R_h, there is a very large velocity gradient near the walls of the pipe except near the edges of the rectangular section. Thus it is possible to apply the boundary layer approximations in those region of large velocity gradient. SHERCLIFF discussed some results of this boundary layer approximations for pipes of cross-section other than the rectangle in reference (*12*).

4. Laminar flow near a stagnation point (*5, 6, 13*). One of the well-known exact solutions of ordinary hydrodynamics of a viscous fluid is the solution of the problem of the flow in the vicinity of a stagnation point. It is possible to find the exact solution in the vicinity of a stagnation point in magnetohydrodynamics. We consider the two dimensional steady flow over an insulated plate situated at $y = 0$ as shown in fig. 6.3. The stagnation point is at the origin. On the surface of the plate, there is an external applied magnetic field of constant strength H_0 in the direction of the y-axis, i. e., perpendicular to the

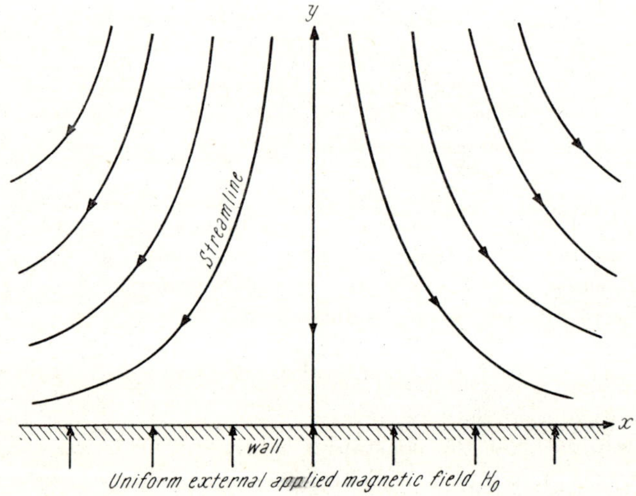

Fig. 6.3. Two-dimensional Flow near a Stagnation Point in Magnetohydrodynamics

plate. The well-known solution for the two-dimensional flow near a stagnation point in the potential flow of an inviscid and incompressible fluid without external magnetic field with the stagnation point at $x = 0$ and $y = 0$ is

$$\begin{gathered} u_i = b_0\, x, \quad v_i = -b_0\, y \\ p_i = p_0 - \frac{\rho}{2} b_0^2 (x^2 + y^2) \end{gathered} \tag{6.25}$$

where b_0 is a constant and p_0 is the stagnation pressure.

For the case of magnetohydrodynamics, we try to find a solution of equation (4.47) in a form similar to equation (6.25). In other words, we generalize the solution of equation (6.25) to the following form:

$$\begin{gathered} u = x\, f'(y), \quad v = -f(y), \quad w = 0 \\ H_x = x\, g'(y), \quad H_y = -g(y), \quad H_z = 0 \\ p = p_0 - p_1(y) - p_2(y)\, x^2, \quad \frac{\partial()}{\partial t} = \frac{\partial()}{\partial z} = 0 \end{gathered} \tag{6.26}$$

where the prime means differentiation with respect to y. Here the equation of continuity (4.47a) and equation (4.9) are automatically satisfied.

Substituting equation (6.26) into equations (4.47b) and (4.47c), we have the following four equations for f, g, p_1 and p_2 which are all functions of y only:

$$f'^2 - f f'' = \frac{2}{\rho} p_2 - \frac{\mu_e}{\rho} g g'' + \frac{\mu}{\rho} f''' \tag{6.27a}$$

$$f f' = \frac{1}{\rho} p_1' - \frac{\mu}{\rho} f'' \tag{6.27b}$$

$$p_2' = \mu_e g' g'' \tag{6.27c}$$

$$-f g' + g f' = \nu_H g'' \tag{6.27d}$$

The boundary conditions for the unknown functions f, g, p_1 and p_2 are that on the plate $u = v = 0$ and $p = p_0 + c x^2$, $H_x = 0$, $H_y = H_0$ and at large distances from the plate $u = b x$. These boundary conditions are based on the no slip conditions of the velocity components. As to the magnetic field, we assume that the plate is insulated, hence there will be no electric current on the wall and its magnetic field strength remains to be the originally external applied magnetic field strength. By boundary conditions of (3.39) and (3.40) without surface current density, we have the continuity of the magnetic field strength of both components $H_x = 0$ and $H_y = H_0 =$ constant at $y = 0$. Hence

$$\begin{gathered} y = 0: f = 0, \quad f' = 0, \quad p_1 = 0, \quad p_2 = C, \quad g = -H_0, \quad g' = 0 \\ y = \infty: \ f' = b \end{gathered} \tag{6.28}$$

where b is a constant in magnetohydrodynamics corresponding to b_0 in the similar problem of ordinary hydrodynamics [cf. (6.25)]. The constant C depends on the applied magnetic field. We may solve equations (6.27a), (6.27c) and (6.27d) for f, g and p_2 with the boundary conditions (6.28), and then substitute the result into equation (6.27b) to get p_1. The numerical factors of the various terms in equation (6.27) may be reduced by the following transformation:

$$\begin{gathered} \eta = \sqrt{\frac{\rho b}{\mu}}\, y, \quad f(y) = \sqrt{\frac{b \mu}{\rho}}\, \phi(\eta), \quad g(y) = \psi(\eta) \cdot \left(\frac{\mu b}{\mu_e}\right)^{\frac{1}{2}} \\ p_1 = \mu b p_1^*(\eta), \quad p_2 = p_2^*(\eta) \rho b^2, \quad \gamma = \frac{\rho \nu_H}{\mu} \end{gathered} \tag{6.29}$$

In terms of the non-dimensional quantities η, ϕ, p_1^*, p_2^* and ψ, equations (6.27a) and (6.27c) gives

$$\phi''' + \phi \phi'' - \phi'^2 = \psi \psi'' - \psi'^2 + K \tag{6.30a}$$

where the constant K is obtained by integrating the inviscid flow equation along the plate and comparing with the pressure equation of (6.26). We obtain

$$K = 1 + \frac{\psi^2(0)}{\gamma} \tag{6.30b}$$

Equation (6.27d) becomes

$$\gamma \psi'' = \phi' \psi - \phi \psi' \tag{6.30c}$$

Finally equation (6.27b) becomes

$$\phi \phi' = p_1{}^{*\prime} - \phi'' \tag{6.30d}$$

The boundary conditions (6.28) become

$$\begin{gathered} \phi(0) = \phi'(0) = 0, \quad p_1{}^*(0) = 0, \quad p_2{}^*(0) = \tfrac{1}{2}\left[1 + \frac{\psi^2(0)}{\gamma}\right] \\ \psi(0) = -\left(\frac{\mu_e H_0{}^2}{\mu b}\right)^{\frac{1}{2}}, \quad \psi'(0) = 0 \end{gathered} \tag{6.31}$$

We have to use numerical integration to calculate $\phi(\eta)$ and $\psi(\eta)$ from equations (6.30a) and (6.30c) with the boundary conditions (6.31). It has been carried out by NEURINGER (*6*) on an analogue computer in reference (*6*). NEURINGER found that the stress on the plate is reduced by the external applied magnetic field. The reduction depends both on the applied magnetic field strength and the electrical conductivity of the plasma.

5. Inviscid flow past a magnetized sphere (*4*). In the last three sections, we consider the cases in which the magnetic field is applied externally. Another interesting class of problem is the case in which the magnetic field originates in the body itself. In other words, we investigate the flow over a magnetized body.

We are going to consider the case of a uniform flow of an inviscid fluid of low electrical conductivity over a weakly magnetized sphere. The magnetic pressure number $R_H = \mu_e H_0{}^2/\rho_0 U^2$ is small. As a result, we may express all the variables in our problem in powers of R_H, i. e.,

$$Q = Q_0 + R_H Q_1 + R_H{}^2 Q_2 + \dots \tag{6.32}$$

where Q represents any one of the variables such as velocity vector, magnetic field strength or pressure. Subscript 0 refers to the values of the basic flow with $R_H = 0$ and subscript 1 refers to the first order perturbation.

Substituting equation (6.32) into equation (4.47), we obtain the following non-dimensional equations in which all the velocity components are in terms of the velocity of the uniform flow U, all lengths are in terms of the radius of the sphere r_0 and the magnetic field is in terms of a reference value H_0:

$$(\nabla \times \vec{q_0}) \times \vec{q_0} = -\nabla (p_0 + \tfrac{1}{2} q_0{}^2) \tag{6.33a}$$

$$\nabla \cdot \vec{q_0} = 0 \tag{6.33b}$$

$$\nabla \times \vec{H_0} = R_\sigma (\vec{q_0} \times \vec{H_0}) \tag{6.34a}$$

$$\nabla \cdot \vec{H_0} = 0 \tag{6.34b}$$

$$(\nabla \times \vec{q_1}) \times \vec{q_0} = -\nabla \overline{p_1} + (\nabla \times \vec{H_0}) \times \vec{H_0} \tag{6.35a}$$

$$\nabla \cdot \vec{q_1} = 0 \tag{6.35b}$$

where the pressure is in terms of $\rho_0 U^2$, with ρ_0 being the density of the fluid and $\overline{p_1} = p_1 + \vec{p_0} \cdot \vec{q_1} \cdot R_\sigma = \sigma \mu_e U L =$ magnetic Reynolds number.

Equation (6.33) shows that the velocity distribution of the basic flow is independent of the magnetic field. For a uniform flow $U = 1$ over a sphere of radius $r_0 = 1$, the potential flow solution of equation (6.33) gives

$$\vec{q}_0 = i_r \left(1 - \frac{1}{r^3}\right) \cos\theta + i_\theta \left(-1 - \frac{1}{r^3}\right) \sin\theta + i_\Phi \cdot 0 \tag{6.36}$$

where i_r, i_θ and i_Φ are respectively the unit vectors in r-, θ- and ϕ-direction of polar coordinates (r, θ, ϕ).

Equations (6.34) give

$$H_r = \frac{1}{r^2 \sin\theta} \cdot \frac{\partial A}{\partial \theta}, \quad H = -\frac{1}{r \sin\theta} \cdot \frac{\partial A}{\partial \theta} \tag{6.37a}$$

and

$$\frac{\partial^2 A}{\partial r^2} + \frac{\sin\theta}{r^2} \cdot \frac{\partial}{\partial\theta}\left(\frac{1}{\sin\theta} \cdot \frac{\partial A}{\partial\theta}\right) = R_\sigma\left[\left(1 - \frac{1}{r^3}\right)\cos\theta \frac{\partial A}{\partial r} - \frac{1}{r}\left(1 + \frac{1}{2r^3}\right)\sin\theta \frac{\partial A}{\partial\theta}\right] \tag{6.37b}$$

For small values of R_σ equation (6.37b) may be solved by method of small perturbation. The perturbation is not regular at infinity. Proper perturbation must be carried out. LUDFORD and MURRAY found that if a dipole is at the center of the sphere, the magnetic field outside the sphere has the following components:

$$H_r = \exp\left[-\tfrac{1}{2} R_\sigma r (1 - \cos\theta)\right]\left\{\frac{2\varkappa}{r^3}\left(1 + \tfrac{1}{2} R_\sigma r\right)\cos\theta + \frac{R_\sigma}{r^4}\left[\left(\tfrac{1}{2}\varkappa r^2 + 3\lambda\right)\cos^2\theta - \left(\tfrac{1}{2}\varkappa r^2 + \lambda\right)\right]\right\} \tag{6.38a}$$

$$H = \exp\left[-\tfrac{1}{2} R_\sigma r (1 - \cos\theta)\right] \cdot \left[\frac{\varkappa}{r^3}\left(1 + \tfrac{1}{2} R_\sigma r\right)\sin\theta - \frac{R_\sigma}{r^4}\left(\tfrac{1}{2}\varkappa r^2 - 2\lambda\right)\sin\theta\cos\theta\right] \tag{6.38b}$$

$$\text{where} \quad \varkappa = \frac{3}{2\mu_e + \mu_e'}, \quad \lambda = \frac{-9\mu_e \mu_e'}{2(2\mu_e + \mu_e') \cdot (3\mu_e + 2\mu_e')}$$

and μ_e' is the permeability of the sphere.

From equation (6.35), we have the vorticity ω_1 equation as follows:

$$\left(r - \frac{1}{r^2}\right)\cos\theta \frac{\partial\bar{\omega}}{\partial r} - \left(1 + \frac{1}{2r^3}\right)\sin\theta \frac{\partial\bar{\omega}}{\partial\theta} = f(r, \theta; R_\sigma) \tag{6.39}$$

$$\text{where} \quad \bar{\omega} = \frac{\omega_1}{r \sin\theta}, \quad \omega_1 = \frac{1}{r}\left(\frac{\partial r q_{\theta 1}}{\partial r} - \frac{\partial q_{r1}}{\partial\theta}\right) \quad \text{and}$$

$$f(r, \theta; R_\sigma) = \exp\left[-R_\sigma r (1 - \cos\theta)\right]\left\{-\frac{3\varkappa^2 R_\sigma}{r^7}(1 + R_\sigma r) \cdot (7\cos\theta + 1) + \right.$$
$$\left. + \frac{\varkappa R_\sigma^2}{r^{11}}\cos\theta\left[\left(-3\varkappa r^5 - 70\lambda r^3 - \tfrac{15}{2} r^2 + 7\lambda\right)\cos^2\theta + \left(9\varkappa r^5 + 6\lambda r^3 + \tfrac{3}{2}\varkappa r^2 + 9\lambda\right)\right]\right\}$$

The left-hand side of equation (6.39) represents a differentiation along the streamlines of the potential flow (6.36). In terms of the coordinates velocity potential ϕ_0 and the streamfunction ψ_0 of the flow such that

$$\phi_0 = r\left(1 + \frac{1}{2r^3}\right)\cos\theta, \quad \psi_0 = r\left(1 - \frac{1}{r^3}\right)^{\frac{1}{2}}\sin\theta \tag{6.40}$$

we may integrate equation (6.39) immediately along a given streamline and obtain

$$\bar{\omega} = \int_{-\infty}^{\phi_0} \frac{r^5 f(r,\theta;R_\sigma)}{[(r^3-1)^2\cos^2\theta + (r^3+\frac{1}{2})^2\sin^2\theta]} \mathrm{d}\phi_0 \tag{6.41}$$

The vorticity is logarithmic infinity at the stagnation point.

The first order streamfunction ψ, satisfies

$$\frac{\partial^2\psi_1}{\partial r^2} + \frac{\sin\theta}{r^2}\cdot\frac{\partial}{\partial\theta}\left(\frac{1}{\sin\theta}\cdot\frac{\partial\psi_1}{\partial\theta}\right) = -r\sin\theta\cdot\omega_1 \tag{6.42}$$

where

$$q_{r1} = \frac{1}{r^2\sin\theta}\cdot\frac{\partial\psi_1}{\partial\theta}, \quad q_{\theta 1} = -\frac{1}{r\sin\theta}\cdot\frac{\partial\psi_1}{\partial r} \tag{6.42a}$$

Equation (6.42) may be solved by a method of series expansion such that

$$\psi_1 = \sum_{n=1}^{\infty} \psi_n(r)\sin^2\theta\, P_n'(\cos\theta) \tag{6.43a}$$

$$f(r,\theta;R_\sigma) = \sum_{n=1}^{\infty} \Omega_n(r)\, P_n'(\cos\theta) \tag{6.43b}$$

$$\omega_1 = \sum_{n=1}^{\infty} \omega_n(r)\sin\theta\, P_n'(\cos\theta) \tag{6.43c}$$

where $P_n(\cos\theta)$ is the Legendre polynomial and prime means first derivative.

The functions ψ_n which vanishes on $r = 1$ are given by

$$\psi_n = \frac{1}{2n+1}\int_1^r \left(\frac{\zeta^{n+2}}{r^n} - \frac{r^{n+1}}{\zeta^{n-1}}\right)\omega_n(\zeta)\,\mathrm{d}\zeta + A_n\left(r^{n+1} - \frac{1}{r^n}\right) \tag{6.44}$$

$$\text{where} \quad A_n = \frac{1}{2n+1}\int_1^{\infty} \frac{\omega_n(\zeta)}{\zeta^{n-1}}\mathrm{d}\zeta$$

The velocity field follows equation (6.42a). In particular

$$q_{\theta 1} = \sum_{n=1}^{\infty} q_n(r)\sin\theta\, P_n'(\cos\theta), \quad q_n(r) = -\frac{\psi_n'(r)}{r} \tag{6.45}$$

For instance

$$q_2(1) = -\tfrac{5}{9}\int_1^{\infty} \Omega_1(\alpha)\,\mathrm{d}\alpha \tag{6.45a}$$

After the velocity and magnetic field are obtained, we may calculate the force on the sphere which consists of two parts: one is due to the pressure of the fluid and the other due to the Maxwell stress.

The drag of a sphere with a dipole at its center due to the pressure of the fluid is

$$D_p = R_H\, 0\,(R_\sigma{}^3) \tag{6.46}$$

and the drag due to Maxwell stress is

$$D_M = R_H [\tfrac{8}{5} \varkappa^2 R_\sigma + 0 (R_\sigma^2)] \tag{6.47}$$

For small magnetic Reynolds number, the first order drag of the sphere is given by the first term in equation (6.47).

6. Boundary layer flow of a fluid of very small electrical conductivity. For aeronautical engineering problems, the magnetic Reynolds number R_σ is usually small [cf. equation (5.28a)]. Under this condition, the induced magnetic field due to the flow may be neglected with respect to the applied magnetic field H_0.

Again we consider only the two dimensional flow. For the velocity, we have only two components u and v with $w = 0$. For the magnetic field we have also only two components H_x and H_y with $H_z = 0$. However the electrical current $\vec{i}$ and the electric field $\vec{E}$ have z-components only. For the case of small σ, we may write (*1*)

$$\begin{aligned} H &= H_0 + \sigma H_1 + \sigma^2 H_2 + \dots \\ E &= E_0 + \sigma E_1 + \sigma^2 E_2 + \dots \end{aligned} \tag{6.48}$$

where the subscript 0 refers to the external applied fields. We shall assume that H_0 has only the y-component with a value of H_{y0}. If we consider the boundary layer flow over a flat plate situated at $y = 0$, in the boundary layer the x-wise velocity component u is much larger than the y-wise velocity component v. We have then

$$\begin{aligned} \vec{u} \times \vec{H}_0 &= \vec{k}_3 (u H_{y0} - v H_{x0}) \cong \vec{k}_3 u H_{y0} \\ \vec{u} \times \vec{H}_1 &= \vec{k}_3 (u H_{y1} - v H_{x1}) \cong \vec{k}_3 u H_{y1} \end{aligned} \tag{6.49}$$

where $\vec{k}_3$ is the unit vector in the z-direction.

From equation (3.22), we have the electrical current density

$$\vec{i} = \sigma (\vec{E} + \mu_e \vec{u} \times \vec{H}) = \sigma (\vec{E}_0 + \mu_e \vec{u} \times \vec{H}_0) + \sigma^2 (\vec{E}_1 + \mu_e \vec{u} \times \vec{H}_1) + \dots \tag{6.50}$$

In the present problem, we have the z-component of $\vec{i}$ only. From equations (6.49) and (6.50), we have

$$i_z = \sigma (E_{z0} + \mu_e u H_{y0}) + \sigma^2 (E_{z1} + \mu_e u H_{y1}) + \dots \tag{6.51}$$

The circuit of the electrical current must be closed. We may assume that the current is taken up by suitable electrodes, connected externally by a circuit with resistance R. If h is the distance between the electrodes (the height of the field in the direction of the z-axis); A, their equivalent area, the current in the external circuit will be $I = A i_z$. We then have the boundary condition

$$h E = - R I = - R A i_z \tag{6.52}$$

Substituting equations (6.48) and (6.51) into equation (6.52), we have

$$E_{z0} + \sigma E_{z1} + \dots = - \beta [(E_{z0} + \mu_e u H_{y0}) + \sigma (E_{z1} + \mu_e u H_{y1}) + \dots] \tag{6.53}$$

where $\beta = R A \sigma / h$.

We will assume that β is small. Now we have the following relations

$$E_{z0} = -\frac{\beta}{1+\beta}\mu_e u H_{y0}$$
$$E_{z1} = -\frac{\beta}{1+\beta}\mu_e u H_{y1} \tag{6.54}$$

etc.

Finally equation (6.51) becomes

$$i_z = \frac{1}{1+\beta}(\sigma \mu_e u H_{y0} + \sigma^2 \mu_e u H_{y1} + \ldots) \tag{6.55}$$

The ponderomotive force in the equations of motion is then

$$\vec{i} \times \vec{B} = -\vec{k}_1 \mu_e i_z H_y + \vec{k}_2 \mu_e i_z H_x$$

where $\vec{k}_1$ and $\vec{k}_2$ are the unit vector in the direction of x and y respectively. The x-wise ponderomotive force is then

$$F_{ex} = \frac{-1}{1+\beta}(\sigma u \mu_e^2 H_{y0}^2 + 2\sigma^2 u \mu_e^2 H_{y0} H_{y1} + \ldots) \tag{6.56}$$

The y-wise ponderomotive force is

$$F_{ey} = \frac{1}{1+\beta}(\sigma^2 u H_{y0} H_{x1} + \ldots) \tag{6.57}$$

For small values of σ and β, we have for first approximation:

$$F_{ex} = -\sigma u \mu_e^2 H_{y0}^2, \quad F_{ey} = 0 \tag{6.58}$$

Equation (6.58) may be used in the equations of motion of the boundary layer flow. For first approximation, we may consider equations (4.47a) and (4.47b) only. The corresponding two dimensional boundary layer equations are then

$$\frac{\partial u}{\partial x} + \frac{\partial v}{\partial y} = 0 \tag{6.59a}$$

$$\frac{\partial u}{\partial t} + u\frac{\partial u}{\partial x} + v\frac{\partial u}{\partial y} = -\frac{1}{\rho}\cdot\frac{\partial p}{\partial x} + \nu\frac{\partial^2 u}{\partial y^2} - \frac{\sigma}{\rho}\mu_e^2 H_{y0}^2 u \tag{6.59b}$$

The only difference of equation (6.59) from ordinary boundary layer equations is the last term in equation (6.59b).

Now we consider the two dimensional steady flow in the boundary layer over a flat without pressure gradient in a uniform stream of velocity u_0 with an external applied magnetic field of constant strength H_{y0}. The conditions of this problem are then

$$y = 0: \quad u = v = 0; \quad \frac{\partial p}{\partial x} = 0, \quad \frac{\partial u}{\partial t} = 0$$
$$y = \infty: \quad u = u_0, \quad v = \frac{\partial u}{\partial y} = 0, \quad \frac{\partial u}{\partial x} = -\frac{\sigma}{\rho}\mu_e^2 H_{y0}^2 = -m_1 \tag{6.60}$$

We may use the similar methods of solution of ordinary boundary layer equation (*6*) to solve equations (6.59) with the conditions (6.60). We introduce a stream function ψ (x, y) such that

$$\frac{\partial \psi}{\partial y} = u, \quad \frac{\partial \psi}{\partial x} = -v \tag{6.61}$$

Equation (6.59a) is automatically satisfied. We write the stream function in the following form:

$$\psi = \sqrt{u_0 \nu x}\,[f_0(\eta) + m x f_2(\eta) + (m x)^2 f_4(\eta) + \ldots] \tag{6.62}$$

$$\text{where } \eta = y\left(\frac{u_0}{\nu x}\right)^{\frac{1}{2}}, \quad m = \frac{m_1}{u_0} \tag{6.63}$$

Substituting equations (6.60), (6.61) and (6.62) into equation (6.59b), we find the following differential equations for f_0, f_2, f_4, etc.

$$2 f_0''' = -f_0'' f_0 \tag{6.64a}$$

$$2 f_2''' = 2 f_0' f_2' - f_0 f_2'' - 3 f_2 f_0'' + 2 f_0' \tag{6.64b}$$

$$2 f_4''' = 4 f_0' f_4' + 2 f_2' f_2' - f_0 f_4'' - 3 f_2 f_2'' - 5 f_4 f_0'' + 2 f_2' \tag{6.64c}$$

etc. The boundary conditions are

$$\begin{aligned} f_0 = f_2 = f_4 = \ldots = 0 \text{ at } \eta = 0 \\ f_0' = f_2' = f_4' = \ldots = 0 \text{ at } \eta = 0 \\ f_0' = 1, \quad f_2' = -1, \quad f_4' = \ldots = 0 \text{ at } \eta = \infty \end{aligned} \tag{6.65}$$

Equation (6.64a) is the well-known Blasius equation whose solution is known. Equations (6.64b), (6.64c), etc. are linear ordinary differential equations for f_2, f_4, etc. respectively. These equations may be integrated numerically. The function f_2 has been calculated by Rossow (*9*), who gave the following formula for the skin friction coefficient for the present case

$$c_f = \frac{0.664 - 1.788\, m x + \ldots}{\sqrt{R_{ex}}} \tag{6.66}$$

If $m = 0$, equation (6.66) reduces to the formula for ordinary boundary flow. With external applied magnetic field, the skin friction coefficient decreases with increase of $m = \sigma \mu_e H_{y0}^2/(\rho u_0)$. $R_{ex} = U_0 x/\nu$ = Reynolds number with respect to distance x along the plate.

7. Boundary layer flow of fluid of large or infinite electrical conductivity. For the case of large electrical conductivity, we cannot neglect the induced magnetic field. We have to solve all the equations (4.47) simultaneously. We consider now a flat plate placed along the direction of a uniform stream of incompressible, viscous, heat-conducting and electrically conducting fluid. For two dimensional steady flow, equations (4.47) in non-dimensional form are

$$u\frac{\partial u}{\partial x} + v\frac{\partial u}{\partial y} - R_H\left(H_x\frac{\partial H_x}{\partial x} + H_y\frac{\partial H_x}{\partial y}\right) = -\frac{\partial}{\partial x}\left(p + R_H\frac{H^2}{2}\right) + \frac{1}{R_e}\left(\frac{\partial^2 u}{\partial x^2} + \frac{\partial^2 u}{\partial y^2}\right) \tag{6.67a}$$

$$u\frac{\partial v}{\partial x}+v\frac{\partial v}{\partial y}-R_H\left(H_x\frac{\partial H_y}{\partial x}+H_y\frac{\partial H_y}{\partial y}\right)=-\frac{\partial}{\partial y}\left(p+R_H\frac{H^2}{2}\right)+\frac{1}{R_e}\left(\frac{\partial^2 v}{\partial x^2}+\frac{\partial^2 v}{\partial y^2}\right) \tag{6.67b}$$

$$u\frac{\partial H_x}{\partial x}+v\frac{\partial H_y}{\partial y}-\left(H_x\frac{\partial u}{\partial x}+H_y\frac{\partial u}{\partial y}\right)=\frac{1}{R_\sigma}\left(\frac{\partial^2 H_x}{\partial x^2}+\frac{\partial^2 H_x}{\partial y^2}\right) \tag{6.67c}$$

$$\frac{\partial u}{\partial x}+\frac{\partial v}{y}=0 \tag{6.67d}$$

$$\frac{\partial H_x}{\partial x}+\frac{\partial H_y}{\partial y}=0 \tag{6.67e}$$

where the non-dimensional quantities u, v; H_x, H_y; and x, y are based on the reference values U, H_0 and L respectively. The non-dimensional pressure is based on ρU^2. The three non-dimensional parameters R_e, R_H and R_σ are defined respectively

$$R_e=\frac{U\,L}{\nu},\quad R_H=\frac{\mu_e H_0^2}{\rho\, U^2},\quad R_\sigma=\sigma\,\mu_e\,U\,L \tag{6.68}$$

and $H^2=H_x^2+H_y^2$.

For boundary layer flow, the Reynolds number R_e is large. The y-wise distance and velocity component are small. Furthermore from equation (6.67e) we can show that the y-component of the magnetic field in the boundary layer is also small. As a result, we may introduce the following new variables:

$$V=v\sqrt{R_e},\quad Y=y\sqrt{R_e},\quad h_y=H_y\sqrt{R_e} \tag{6.69}$$

Substituting the relations (6.69) into equations (6.67) and neglecting the terms of $1/R_e$ or smaller, we have the boundary layer equations of magnetohydrodynamics as follows:

$$u\frac{\partial u}{\partial x}+V\frac{\partial u}{\partial Y}-R_H\left(H_x\frac{\partial H_x}{\partial x}+h_y\frac{\partial H_x}{\partial Y}\right)=-\frac{\partial}{\partial x}\left(p+R_H\frac{H_x^2}{2}\right)+\frac{\partial^2 u}{\partial Y^2} \tag{6.70a}$$

$$\frac{\partial}{\partial Y}\left(p+R_H\frac{H_x^2}{2}\right)=0 \tag{6.70b}$$

$$\frac{R_\sigma}{R_e}\left(u\frac{\partial H_x}{\partial x}+V\frac{\partial H_x}{\partial Y}-H_x\frac{\partial u}{\partial x}-h_y\frac{\partial u}{\partial Y}\right)=\frac{\partial^2 H_x}{\partial Y^2} \tag{6.70c}$$

$$\frac{\partial u}{\partial x}+\frac{\partial V}{\partial Y}=0 \tag{6.70d}$$

$$\frac{\partial H_x}{\partial x}+\frac{\partial h_y}{\partial Y}=0 \tag{6.70e}$$

From equation (6.70b), we have

$$p+R_H\frac{H_x^2}{2}=F(x) \tag{6.71}$$

The arbitrary function $F(x)$ is assumed known in the study of the boundary layer flow. At the edge of the boundary layer, we assume

$$\frac{\partial H_x}{\partial Y} = \frac{\partial u}{\partial Y} = 0 \tag{6.72}$$

Hence

$$\frac{\partial}{\partial x}\left(p + R_H \frac{H_x^2}{2}\right) = -u_\infty \frac{\partial u_\infty}{\partial x} + R_H H_{x\infty} \frac{\partial H_{x\infty}}{\partial x} \tag{6.73}$$

where the subscript ∞ represents the value at the edge of the boundary layer, i. e., the value of the inviscid solution.

We may introduce streamfunctions ψ and ϕ for the velocity field and the magnetic field respectively as follows:

$$u = \frac{\partial \psi}{\partial Y}, \quad V = -\frac{\partial \psi}{\partial x} \tag{6.74}$$

$$H_x = \frac{\partial \phi}{\partial Y}, \quad h_y = -\frac{\partial \phi}{\partial x} \tag{6.75}$$

Equations (6.70a) and (6.70c) in terms of ψ and ϕ become respectively

$$\frac{\partial \psi}{\partial Y}\cdot\frac{\partial^2 \psi}{\partial x \partial Y} - \frac{\partial \psi}{\partial x}\cdot\frac{\partial^2 \psi}{\partial Y^2} - R_H\left(\frac{\partial \phi}{\partial Y}\cdot\frac{\partial^2 \phi}{\partial x \partial Y} - \frac{\partial \phi}{\partial x}\cdot\frac{\partial^2 \phi}{\partial Y^2}\right) = -\frac{\mathrm{d} F}{\mathrm{d} x} + \frac{\partial^3 \psi}{\partial Y^3} \tag{6.76a}$$

$$\frac{\partial^3 \phi}{\partial Y^3} = \frac{R_\sigma}{R_e}\left(\frac{\partial \psi}{\partial Y}\cdot\frac{\partial^2 \phi}{\partial x \partial Y} - \frac{\partial \psi}{\partial x}\cdot\frac{\partial^2 \phi}{\partial Y^2} - \frac{\partial \phi}{\partial Y}\cdot\frac{\partial^2 \psi}{\partial x \partial Y} + \frac{\partial \phi}{\partial x}\cdot\frac{\partial^2 \psi}{\partial Y^2}\right) \tag{6.76b}$$

The boundary layer flow problem is to find ψ and ϕ for various boundary conditions.

We shall consider a simple problem in which $u = U =$ constant and $H_x = H_0 =$ constant. Furthermore at the plate $h_y = 0$. If the magnetic Reynolds number R_σ is infinite, the solution $\psi = \phi$ satisfies our problem and equation (6.76b) is automatically satisfied. Equation (6.76a) becomes

$$(1 - R_H)\cdot\left(\frac{\partial \psi}{\partial Y}\cdot\frac{\partial^2 \psi}{\partial x \partial Y} - \frac{\partial \psi}{\partial x}\cdot\frac{\partial^2 \psi}{\partial Y^2}\right) = \frac{\partial^3 \psi}{\partial Y^3} \tag{6.77}$$

Except the factor $(1 - R_H)$, equation (6.77) is identical to the Blasius equation of ordinary boundary layer equation. Hence Blasius solution with some modification by the factor $(1 - R_H)$ satisfies our problem too. However, the solution breaks down at $R_H = 1$. The reason is that at $R_H = 1$, the inertial forces are cancelled by the electromagnetic force. As a result, the boundary layer approximations no longer hold true.

When R_H is smaller than unity, the skin friction decreases as R_H increases similar to the case of low magnetic Reynolds numbers.

When R_H is larger than unity, we have upstream wake effect.

8. Unsteady motion of a plate. Rayleigh's problem (*9, 14, 15*). We consider the case that an infinite flat insulated plate sets impulsively into uniform motion with velocity U in its own plane in the presence of a transverse uniform magnetic field of strength H_0 as shown in fig. 6.4. The conditions of this problem are

$$\begin{aligned} &u = u(y, t), \quad v = w = 0, \quad p = \text{constant} \\ &H_x = H_x(y, t), \quad H_y = H_0 = \text{constant}, \quad H_z = 0 \end{aligned} \tag{6.78}$$

Now we are going to consider the functions $u(y, t)$ and $H_x(y, t)$ only. Equation (4.47) gives the following differential equations for u and H_x:

$$\rho \frac{\partial u}{\partial t} - \mu_e H_0 \frac{\partial H_x}{\partial y} = \mu \frac{\partial^2 u}{\partial y^2} \tag{6.79a}$$

$$\frac{\partial H_x}{\partial t} - H_0 \frac{\partial u}{\partial y} = + \nu_H \frac{\partial^2 H_x}{\partial y^2} \tag{6.79b}$$

The initial conditions are

$$t = 0: \quad u = 0, \quad H_x = 0 \tag{6.80}$$

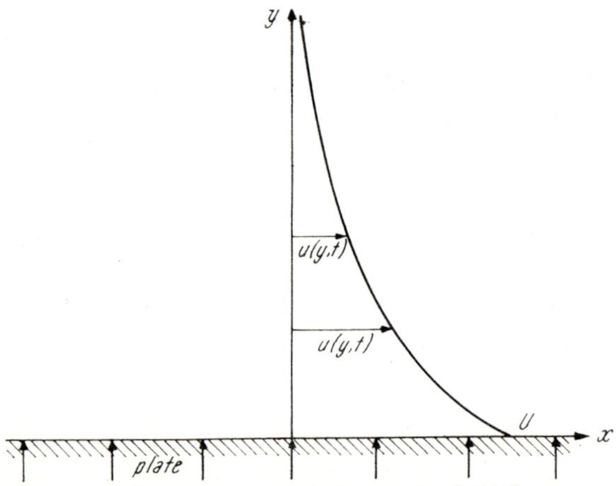

Fig. 6.4. Rayleigh problem in Magnetohydrodynamics

The boundary conditions are

$$\begin{aligned} y = 0: &\quad u = U, \quad H_x = 0 \\ y = \infty: &\quad u = 0, \quad H_x = 0 \end{aligned} \tag{6.81}$$

The solution of equation (6.79) with the conditions (6.80) and (6.81) may be obtained by the method of Laplace transform. But the Laplace transform solution is very complicated and its exact inversion would be extremely difficult. Yang discussed this solution in reference *14* and he also gave explicit solutions for the limiting cases of very weak magnetic field and very strong magnetic field. Here we shall give an approximate solution for the case of very small electric conductivity σ. In this case we may replace equation (6.79b) by the following relation [cf. (6.55)]:

$$\frac{\partial H_x}{\partial y} = - \left(\frac{1}{\nu_H}\right) H_0 u \tag{6.82}$$

Equation (6.79a) with relation (6.82) becomes

$$\frac{\partial u}{\partial t} + m_1 u = \nu \frac{\partial^2 u}{\partial y^2} \tag{6.83}$$

where $m_1 = + [\sigma/\rho] \mu_e^2 H_0^2$.

The boundary conditions for u are still those given in equation (6.80) and (6.81).

We solve equation (6.83) by the Laplace transform. The Laplace transform of the velocity u is defined as (*9*)

$$\bar{u} = \int_0^\infty e^{-st}\, u \,\mathrm{d}t \tag{6.84}$$

The transformed equation of (6.83) is then

$$\nu \frac{\mathrm{d}^2 \bar{u}}{\mathrm{d}y^2} = \bar{u}\,(s + m_1) \tag{6.85}$$

The solution of equation (6.85) with the corresponding conditions to (6.80) and (6.81) is

$$\bar{u} = \frac{U}{(s + m_1) - m_1}\, e^{-\left[(s + m_1)^{1/2} \frac{y}{\sqrt{\nu}}\right]} \tag{6.86}$$

The inversion of equation (6.86) gives the velocity distribution[1]

$$u\,(y, t) = \frac{U}{2}\left[e^{-y\sqrt{\frac{m_1}{\nu}}}\, erfc\left(\frac{y}{2\sqrt{\nu t}} - m_1 t\right) + e^{+y\sqrt{\frac{m_1}{\nu}}}\, erfc\left(\frac{y}{2\sqrt{\nu t}} + m_1 t\right)\right] \tag{6.87}$$

where the symbol *erfc* denotes the complementary error function defined as $1 - erf$. When $m_1 = 0$, equation (6.87) reduces to the solution of ordinary hydrodynamics.

9. The temperature distribution in magnetohydrodynamics. After the velocity distribution and the magnetic field strength distributions are obtained, the temperature distribution may be calculated from the energy equation, e. g., equation (4.48) or (4.49). Since the energy equation is linear in terms of the temperature, the calculation of the temperature distribution can be easily carried out. Sometimes numerical integration must be used because of the variable coefficients of this equation due to the distributions of velocity.

We shall consider a simple case of the boundary layer flow corresponding to equation (6.59). The energy equation of a steady two-dimensional boundary layer flow without pressure gradient is (*9*)

$$\rho\, u\, c_p \frac{\partial T}{\partial x} + \rho\, v\, c_p \frac{\partial T}{\partial y} - \sigma\, \mu_e^2\, H_0^2\, u^2 = \varkappa \frac{\partial^2 T}{\partial y^2} + \mu \left(\frac{\partial\, u^2}{\partial\, y}\right) \tag{6.88}$$

If $P_r = c_p\, \mu/\varkappa = 1$, equation (6.88) becomes

$$u \frac{\partial h_0}{\partial x} + v \frac{\partial h_0}{\partial y} = \nu \frac{\partial^2 h_0}{\partial y^2} \tag{6.89}$$

where

$$h_0 = c_p\, T + \frac{u^2}{2} \tag{6.90}$$

It is interesting to notice that under the present approximation, the Busemann

[1] Chang and Yen in reference *15* showed that for finite electrical conductivity the solution of this problem differs appreciably from this limiting solution.

and Crocco's relation (6.89) of ordinary boundary layer flow still holds in the present case. For insulated plate with respect to heat transfer, we have

$$h_0 = c_p T + \frac{u^2}{2} = \text{constant} \tag{6.91}$$

This is the well-known Busemann's relation. However in the present case with heat transfer, the Crocco's relation of ordinary boundary flow does not hold true because equation (6.89) is not the same as the equation of motion (6.59b) with $\partial u/\partial t = \partial p/\partial x = 0$. However, we may use the well known Pohlhausen method (*7*) of solution of temperature distribution of boundary layer flow in the present case.

References

1. Burgers, J. M.: Magnetogasdynamic Interaction. Cornell University Report, 1952.

2. Hartmann, J.: Hydrodynamics I. Theory of the Laminar Flow of an Electrically Conductive Liquid in a Homogeneous Magnetic Field. Kgl. Danske Videnskabernes Selskab, Math.-Fys. Med., vol. 15, No. 6, Copenhagen, 1937.

3. Hartmann, J., and F. Lazarus: Hg-Dynamics II–Experimental Investigations on the Flow of Mercury in a Homogeneous Magnetic Field. Kgl. Danske Videnskabernes Selskab, Math.-Fys. Med., vol. 14, No. 7, Copenhagen, 1937.

4. Ludford, G. G. S., and J. D. Murray: On the Flow of a Conducting Fluid past a Magnetized Sphere. Jour. Fluid Mach., vol. 7, pt. 4, pp. 516–528, 1960.

5. Meyer, R. C.: On Reducing Aerodynamic Heat-transfer Rates by Magnetohydrodynamic Techniques. Jour. Aero. Sci., vol. 25, No. 9, Sept. 1958, pp. 561–566.

6. Neuringer, J. L.: Two Dimensional Flow in the Vicinity of the Stagnation Point of an Incompressible Viscous, Electrically Conducting Fluid in the Presence of a Magnetic Field. Republic Aviation Corp. Report, June 12, 1957.

7. Pai, S. I.: Viscous Flow Theory. I–Laminar Flow 1956 and II–Turbulent Flow 1957. D. Van Nostrand Co., Princeton, N. J.

8. Pai, S. I.: Laminar Flow of an Electrically Conducting Incompressible Fluid in a Circular Pipe. Jour. App. Phys., vol. 25, No. 9, Sept. 1954, pp. 1205–1207.

9. Rossow, V. J.: On Flow of Electrically Conducting Fluids over a Flat Plate in the Presence of a Transverse Magnetic Field. N. A. C. A. Tech., Note 3971, May 1957.

10. Resler, E. L. Jr., and W. R. Sears: The Prospects for Magneto-aerodynamics. Jour. Aero. Sci., vol. 25, No. 4, April 1958, pp. 235–245.

11. Shercliff, J. A.: Steady Motion of Conducting Fluids in Pipes under Transverse Magnetic Fields. Proc. Cambridge Phil. Soc., vol. 49, pt. 1, Jan. 1953, pp. 136–144.

12. Watson, G. N.: Theory of Bessel Functions. Cambridge Univ. Press, 1944.

13. Wu, C. S., and W. D. Hayes: Axisymmetric Stagnate Flow of a Viscous and Electrically Conducting Fluid near the Blunt Nose of a Spinning Body with Presence of Magnetic Field. Part I: Exact Solution of Incompressible and Constant Properties Model. Princeton University report No. 431, April 1958.

14. Yang, H. T.: Rayleigh Problem in a Transverse Magnetic Field. Report of Institute for Fluid Dynamics and Applied Mathematics, University of Maryland, 1958.

15. Yen, J. T., and C. C. Chang: On Rayleigh's Problem in Magnetohydrodynamics. Report of Aero. Eng. Dept., University of Minnesota, Dec. 1958.

Chapter VII

Stability of Magnetogasdynamic Flows

1. Introduction. One of the most interesting problems in fluid dynamics is the problem of the stability of a laminar flow and its transition to turbulence. It will be of interest to see what effect the magnetic field has on the stability of laminar flow in magnetogasdynamics. In this chapter we are going to discuss some of the stability problems in magnetogasdynamics.

The most extensive discussion of hydrodynamic stability seems to be the treatment of parallel flows by attempting to solve the eigenvalue problem associated with the linearized equations governing the disturbance. In § 2, we shall discuss these linearized equations of disturbances for an incompressible, electrically conducting fluid under external applied magnetic field. In § 3, the stability of the flow in a two-dimensional channel under a coplanar magnetic field will be discussed while in § 4 the corresponding problem but under a transverse magnetic field will be studied. In § 5, we discuss the stability of a boundary layer flow for the case of small electrical conductivity and also the stability of a laminar jet mixing problem.

The most successful case of the analysis of the stability problem of a laminar flow appeared to be Taylor's treatment of Couette flow between concentric cylinders. In § 6, we shall discuss the corresponding case in magnetohydrodynamics.

Another important type of instability of flow problem is the thermal instability which leads to the onset of convection in a fluid heated from below. We shall discuss the thermal instability in a magnetic field in § 7.

The pinch effect is a very important concept in the controlled fusion research. In § 8, we shall show the instability of the simple pinch effect. The stabilization of the pinch effect is one of the most important problem which has to be solved before the controlled fusion reaction can be put into practical uses.

An experiment of stability of flow in magnetohydrodynamics is briefly discussed in § 9.

2. Stability equations for parallel flow in magnetohydrodynamics. The basic flow in our investigation is a two-dimensional steady parallel flow with the following values of velocity and magnetic field strength:

$$\begin{gathered} u = U(y), \quad v = w = 0, \quad p = A\,x + B + C(y) \\ H_x = H_x(y), \quad H_y = H_0, \quad H_z = 0 \end{gathered} \tag{7.1}$$

where A, B and H_0 are constants. $U(y)$, $C(y)$ and $H_x(y)$ are functions of y only. The basic flow is perturbed by a small disturbance. In the resultant disturbed motion, we have

$$\begin{gathered} u = U(y) + u_x, \quad v = u_y, \quad w = u_z, \quad p = Ax + B + C(y) + p' \\ H_x = H_x(y) + h_x, \quad H_y = H_0 + h_y, \quad H_z = h_z \end{gathered} \tag{7.2}$$

where u_x, u_y, u_z, p', h_x, h_y, and h_z are the perturbed quantities which are assumed to be small. If we substitute the quantities of (7.2) into equations (4.47) and (4.9) and neglect the quadratic terms of the perturbed quantities, we have the equations for the disturbances as follows:

$$\begin{gathered} \frac{\partial u_x}{\partial t} + U\frac{\partial u_x}{\partial x} + u_y\frac{\partial U}{\partial y} = \\ = \frac{\mu_e}{\rho}\left(H_x\frac{\partial h_x}{\partial x} + H_0\frac{\partial h_x}{\partial y} + h_y\frac{\partial H_x}{\partial y}\right) - \frac{1}{\rho}\cdot\frac{\partial p'}{\partial x} - \frac{\mu_e}{\rho}\cdot\frac{\partial}{\partial x}(H_x h_x + H_0 h_y) + \nu\nabla^2 u_x \\ \frac{\partial u_y}{\partial t} + U\frac{\partial u_y}{\partial x} = \frac{\mu_e}{\rho}\left(H_x\frac{\partial h_y}{\partial x} + H_0\frac{\partial h_y}{\partial y}\right) - \frac{1}{\rho}\cdot\frac{\partial p'}{\partial y} - \frac{\mu_e}{\rho}\cdot\frac{\partial}{\partial y}(H_x h_x + H_0 h_y) + \nu\nabla^2 u_y \\ \frac{\partial u_z}{\partial t} + U\frac{\partial u_z}{\partial x} = \frac{\mu_e}{\rho}\left(H_x\frac{\partial h_z}{\partial x} + H_0\frac{\partial h_z}{\partial y}\right) - \frac{1}{\rho}\cdot\frac{\partial p'}{\partial z} - \frac{\mu_e}{\rho}\cdot\frac{\partial}{\partial z}(H_x h_x + H_0 h_y) + \nu\nabla^2 u_z \\ \frac{\partial h_x}{\partial t} + U\frac{\partial h_x}{\partial x} + u_y\frac{\partial H_x}{\partial y} - H_x\frac{\partial u_x}{\partial x} - H_0\frac{\partial u_x}{\partial y} - h_y\frac{\partial U}{\partial y} = \nu_H\nabla^2 h_x \\ \frac{\partial h_y}{\partial t} + U\frac{\partial h_y}{\partial x} - H_x\frac{\partial u_y}{\partial x} - H_0\frac{\partial u_y}{\partial y} = \nu_H\nabla^2 h_y \\ \frac{\partial h_z}{\partial t} + U\frac{\partial h_z}{\partial x} - H_x\frac{\partial u_z}{\partial x} - H_0\frac{\partial u_z}{\partial y} = \nu_H\nabla^2 h_z \\ \frac{\partial u_x}{\partial x} + \frac{\partial u_y}{\partial y} + \frac{\partial u_z}{\partial z} = 0 \\ \frac{\partial h_x}{\partial x} + \frac{\partial h_y}{\partial y} + \frac{\partial h_z}{\partial z} = 0 \\ \text{where} \quad \nabla^2 = \frac{\partial^2}{\partial x^2} + \frac{\partial^2}{\partial y^2} + \frac{\partial^2}{\partial z^2}. \end{gathered} \tag{7.3}$$

Now we assume that the disturbances may be analyzed into an infinite number of principal disturbances, each of which involves the time only through a single exponential factor (*12*). The disturbance is also supposed to be analyzed into constituents which are simple harmonic functions both of x and of z. Then the equations being linear, we need consider only a single constituent. Thus we take u_x, u_y, u_z, p', h_x, h_y, and h_z in the following forms:

$$\begin{aligned} u_x &= u_1(y)\exp i(\alpha x + \alpha_1 z - \beta t) \\ u_y &= v_1(y)\exp i(\alpha x + \alpha_1 z - \beta t) \\ u_z &= w_1(y)\exp i(\alpha x + \alpha_1 z - \beta t) \\ p' &= p_1(y)\exp i(\alpha x + \alpha_1 z - \beta t) \\ h_x &= h_1(y)\exp i(\alpha x + \alpha_1 z - \beta t) \\ h_y &= h_2(y)\exp i(\alpha x + \alpha_1 z - \beta t) \\ h_z &= h_3(y)\exp i(\alpha x + \alpha_1 z - \beta t) \end{aligned} \tag{7.4}$$

where either the real or the imaginary part may be taken at the end to give u_x, u_y, etc. real values. Here α and α_1 are positive real numbers which are the wave numbers of the disturbance; β is the phase velocity of the disturbance which may be complex, i. e., $\beta = \beta_R + i\,\beta_i$. The value of β is to be found to satisfy all the conditions of any particular problem. If the imaginary part of β is positive, the motion is unstable while if the imaginary part of β is negative, the motion is stable.

Substituting equation (7.4) into equation (7.3), we have

$$u_1'' + \left(i\frac{\beta}{\nu} - \alpha^2 - \alpha_1^2\right) u_1 = i\frac{\alpha}{\rho\nu}[p_1 + \mu_e(H_x h_1 + H_0 h_2)] + \\ + \frac{v_1}{\nu}U' + i\frac{\alpha}{\nu}\left(U u_1 - \frac{\mu_e}{\rho}H_x h_1\right) - \frac{\mu_e}{\nu}H_0 h_1' - \frac{\mu_e}{\rho\nu}h_2 H_x' \tag{7.5}$$

$$v_1'' + \left(i\frac{\beta}{\nu} - \alpha^2 - \alpha_1^2\right) v_1 = \\ = \frac{1}{\rho\nu}[p_1' + \mu_e(H_x h_1 + H_0 h_2)'] + i\frac{\alpha}{\nu}\left(U v_1 - \frac{\mu_e}{\rho}H_x h_2\right) - \frac{\mu_e}{\rho\nu}H_0 h_2' \tag{7.6}$$

$$w_1'' + \left(i\frac{\beta}{\nu} - \alpha^2 - \alpha_1^2\right) w_1 = \\ = i\frac{\alpha_1}{\rho\nu}[p_1 + \mu_e(H_x h_1 + H_0 h_2)] + i\frac{\alpha}{\nu}\left(U w_1 - \frac{\mu_e}{\rho}H_x h_3\right) - \frac{\mu_e}{\rho\nu}H_0 h_3' \tag{7.7}$$

$$h_1'' + \left(i\frac{\beta}{\nu_H} - \alpha^2 - \alpha_1^2\right) h_1 = i\frac{\alpha}{\nu_H}(U h_1 - H_x u_1) + \frac{1}{\nu_H}(v_1 H_x' - H_0 u_1' - h_2 U') \tag{7.8}$$

$$h_2'' + \left(i\frac{\beta}{\nu_H} - \alpha^2 - \alpha_1^2\right) h_2 = i\frac{\alpha}{\nu_H}(U h_2 - H_x v_1) + \frac{1}{\nu_H}(-H_0 v_1') \tag{7.9}$$

$$h_3'' + \left(i\frac{\beta}{\nu_H} - \alpha^2 - \alpha_1^2\right) h_3 = i\frac{\alpha}{\nu_H}(U h_3 - H_x w_1) + \frac{1}{\nu_H}(-H_0 w_1') \tag{7.10}$$

$$i\,\alpha\, u_1 + v_1' + i\,\alpha_1 w_1 = 0 \tag{7.11}$$

$$i\,\alpha\, h_1 + h_2' + i\,\alpha_1 h_3 = 0 \tag{7.12}$$

where prime means differentiation with respect to y.

Differentiating equation (7.6) with respect to y, multiplying equation (7.5) by $i\,\alpha$ and (7.7) by $i\,\alpha_1$ and adding the resulting equations, we have with the help of equations (7.11) and (7.12)

$$\frac{1}{\rho\nu}\left(\frac{d^2}{d\,y^2} - \alpha^2 - \alpha_1^2\right)\cdot[p_1 + \mu_e(H_x h_1 + H_0 h_2)] + 2\,i\frac{\alpha}{\nu}\left(v_1 U' - \frac{\mu_e}{\rho}h_2 H_x'\right) = 0 \tag{7.13}$$

Differentiating equation (7.13) with respect to y we have

$$\frac{1}{\rho\nu}D[p_1 + \mu_e(H_x h_1 + H_0 h_2)]' + 2\,i\frac{\alpha}{\nu}\left(v_1 U' - \frac{\mu_e}{\rho}h_2 H_x'\right)' = 0 \tag{7.14}$$

where $D = d^2/(d\,y^2) - \alpha^2 - \alpha_1^2$. Operating D on equation (7.6) we have

$$D D_1 v_1 = \frac{1}{\rho \nu} D [p_1 + \mu_e (H_x h_1 + H_0 h_2)]' + i \frac{\alpha}{\nu} D \left(U v_1 - \frac{\mu_e}{\rho} H_x h_2 \right) - \frac{\mu_e H_0}{\rho \nu} D h_2' \tag{7.15}$$

where $D_1 = \mathrm{d}^2/(\mathrm{d}\, y^2) + i\, \beta/\nu - \alpha^2 - \alpha_1^2$.

Adding equations (7.14) and (7.15) we have

$$D D_1 v_1 - i \frac{\alpha}{\nu} U\, D\, v_1 + i \frac{\alpha}{\nu} v_1 U'' = \frac{i \alpha \mu_e}{\nu} h_2 H_x'' - \frac{\mu_e}{\rho \nu} (H_0 D h_2' + i \alpha H_x D h_2) \tag{7.16}$$

Equation (7.9) may be written as follows:

$$\left(D_2 - i \frac{\alpha}{\nu_H} U \right) h_2 = - \frac{1}{\nu_H} \left(i \alpha H_x + H_0 \frac{\mathrm{d}}{\mathrm{d}\, y} \right) v_1 \tag{7.17}$$

where $D_2 = \mathrm{d}^2/(\mathrm{d}\, y^2) + i\, \beta/\nu_H - \alpha^2 - \alpha_1^2$.

Equations (7.16) and (7.17) are the fundamental equations for the stability of parallel flow (7.1) in magnetohydrodynamics in the presence of magnetic field. If there is no magnetic field, equation (7.17) is identical to zero and equation (7.16) becomes the ordinary stability equation of laminar parallel flow with three dimensional disturbances.

In stability analysis, we usually consider only two dimensional disturbances in which we put $w = h_3 = 0$ and $\alpha_1 = 0$. Equations (7.16) and (7.17) reduce to the following form:

$$D_0 D_{10} v_1 - i \frac{\alpha}{\nu} U D_0 v_1 + i \frac{\alpha}{\nu} v_1 U'' = i \frac{\alpha \mu_e}{\nu} h_2 H_x'' - \frac{\mu_e}{\rho \nu} (H_0 D_0 h_2' + i H_x D_0 h_2) \tag{7.18}$$

$$\left(D_{20} - i \frac{\alpha}{\nu_H} U \right) h_2 = - \frac{1}{\nu_H} \left(i \alpha H_x + H_0 \frac{\mathrm{d}}{\mathrm{d}\, y} \right) v_1 \tag{7.19}$$

where subscript 0 refers to the operators with $\alpha_1 = 0$.

For stability problems, we should solve the eigenvalue problems of equations (7.18) and (7.19) with proper boundary conditions. The exact solutions for arbitrary values of ν and ν_H have not been carried out. Only the cases where the magnetic Reynolds number R_σ is much smaller than the Reynolds number R_e, i. e., $R_e/R_\sigma = \nu_H/\nu$ is very large, have been carried out. We shall discuss two of such cases in the next two sections.

3. Stability of laminar flow between parallel planes in the presence of a coplanar magnetic field (*16*). We consider the stability of laminar flow between two parallel plates at $y = \pm L$ un der a constant pressure gradient and a coplanar magnetic field. The basic flow has the following values:

$$U(y) = U_0 \left(1 - \frac{y^2}{L^2} \right), \quad p = A\, x + B, \quad H_x = H_1 = \text{constant}, \quad H_0 = 0 \tag{7.20}$$

where $U_0 = - L^2/(2\, \mu)\, A$. Since the magnetic field is coplanar to the velocity, the velocity distribution is not affected by the magnetic field. We thus have the parabolic velocity distribution of ordinary plane Poiseuille flow.

We shall consider the case of very small electrical conductivity σ or equivalently a very large value of ν_H. Under this condition, equation (7.19) with the values of basic flow (7.20) may be replaced by the following approximate relation:

$$(h_2'' - \alpha^2 h_2) = -\frac{i\,\alpha\,H_1\,v_1}{\nu_H} \tag{7.21}$$

Substituting equations (7.20) and (7.21) into equation (7.18), we have

$$(U - c)\cdot(v_1'' - \alpha^2 v_1) - v_1\,U'' + i\,\frac{\mu_e\,H_1^2\,\alpha}{\rho\,\nu_H}\,v_1 = -\,i\,\frac{\nu}{\alpha}\,(v_1'''' - 2\,\alpha^2\,v_1'' + \alpha^4\,v_1) \tag{7.22}$$

where $c = \beta/\alpha$. Now we introduce the following non-dimensional quantities:

$$U^* = \frac{U}{U_0}, \quad c^* = \frac{c}{U_0}, \quad v_1^* = \frac{v_1}{U_0}, \quad y^* = \frac{y}{L}, \quad \alpha^* = \alpha\,L$$
$$R_H = \frac{\mu_e\,H_1^2}{\rho\,U_0^2}, \quad R_\sigma = \frac{U_0\,L}{\nu_H}, \quad R_e = \frac{U_0\,L}{\nu} \tag{7.23}$$

Equation (7.22) in the non-dimensional form becomes

$$(U^* - c^*)\cdot(v_1^{*''} - \alpha^{*2}\,v_1^*) - v_1^*\,U^{*''} + i\,R_H\,R_\sigma\,\alpha^*\,v_1^* = -\frac{i}{\alpha^*\,R_e}\,(v_1^{*''''} - 2\,\alpha^{*2}\,v_1^{*''} + \alpha^{*4}\,v_1^*) \tag{7.24}$$

where prime refers to differentiation with respect to y^*.

The only difference between equation (7.24) and the classical Orr-Sommerfeld equation of hydrodynamical stability is the additional term $i\,R_H\,R_\sigma\,\alpha^*\,V_1^*$. Since equation (7.24) is of the same order as the Orr-Sommerfeld equation, we may use a similar method to solve equation (7.24) as that used in solving the Orr-Sommerfeld equation; and in particular the possible eigenvalues of α^*, c^* and R_e are uniquely determined by the boundary conditions on the velocity alone, i. e.,

$$y^* = \pm 1:\ v_1^* = v_1^{*\prime} = 0 \tag{7.25}$$

This calculation has been carried out by STUART in reference *16*. STUART found that the magnetic field has a stabilizing effect on the flow. For a small value of $\alpha^*\,R_H\,R_\sigma$, the neutral stability curve has essentially the same form as that of the ordinary hydrodynamical stability problem (*9*) but lies inside the neutral stability curve of ordinary hydrodynamics. For a large value of $\alpha^*\,R_H\,R_\sigma$ he found that the neutral stability curve forms a loop and the flow is stable for very large Reynolds numbers. However, he pointed out that the validity of the approximation used in the calculation for such large value of $\alpha^*\,R_H\,R_\sigma$ is questionable. Further study based on the original equations (7.18) and (7.19) should be made to clarify this point.

The Squire's theorem of the stability of ordinary hydrodynamics holds here. In other words, if we put in the stability equations of three dimensional disturbances, i. e., equations (7.16) and (7.17):

$$\alpha_0 = (\alpha^2 + \alpha_1^2)^{\frac{1}{2}}, \quad \alpha\,U_3 = \alpha_0\,U_2 \tag{7.26}$$

where U_3 is the characteristic velocity in the case of three dimensional disturbance and U_2, that of two dimensional disturbance, equations (7.16) and (7.17) reduce to the same form as equations (7.18) and (7.19) respectively (*11*). However for the coplanar magnetic field case where $H_x = H_1$ and $H_z = H_3 =$ constant $\neq 0$, the Squire's theorem does not hold in magnetohydrodynamics.

4. Stability of laminar flow between parallel planes under a transverse magnetic field (*10*). In the present case, the basic flow is that discussed in chapter VI, § 2. Here the velocity distribution of the basic flow depends on the strength of the magnetic field.

We shall also consider only the case of very small electrical conductivity σ or very large values of ν_H. Under this condition, the induced magnetic field $H_x(y)$ is negligible in comparison to the applied magnetic field H_0. Therefore, equation (7.19) may be replaced by the following relation:

$$\nu_H (h_2'' - \alpha^2 h_2) = - H_0 v_1' \tag{7.27}$$

Substituting equation (7.27) into equation (7.18), we have for this approximation the stability equation:

$$(U - c) \cdot (v_1'' - \alpha^2 v_1) - v_1 U'' - i \frac{\mu_e H_0^2}{\rho \nu_H \alpha} v_1'' = - i \frac{\nu}{\alpha} (v_1'''' - 2 \alpha^2 v_1'' + \alpha^4 v_1) \tag{7.28}$$

In terms of non-dimensional variables of (7.23) except that now $R_H = \mu_e H_0^2/(\rho U_0^2)$ equation (7.28) becomes

$$\begin{aligned}(U^* - c^*) \cdot (v_1^{*''} - \alpha^{*2} v_1^*) - v_1^* U^{*''} - i \frac{R_H R_\sigma}{\alpha^*} v_1^{*''} = \\ = - \frac{i}{\alpha^* R_e} (v_1^{*''''} - 2 \alpha^{*2} v_1^{*''} + \alpha^{*4} v_1^*)\end{aligned} \tag{7.29}$$

Again equation (7.29) has only one additional term $- i (R_H R_\sigma/\alpha^*) v_1^{*''}$ over the Orr-Sommerfeld equation. Equation (7.29) is also of the fourth order. Thus we may determine the possible eigenvalues of α^*, c^* and R_e^* by the boundary conditions (7.25). The calculations have been carried out by Lock in reference *10*. Lock found that the magnetic field has a stabilizing effect on the flow too. The neutral stability curves have a form similar to that of ordinary hydrodynamics; but as $R_H R_\sigma$ increases, the curves shift in the direction of increasing α^* and R_e. It was also found that the principal effect of the magnetic field in the present case is to modify the velocity distribution of the basic flow. Lock showed that for very large values of $R_H R_\sigma$, the flow becomes completely stable; but the validity of his analysis for very large $R_H R_\sigma$ is questionable. Further study based on the original equations (7.18) and (7.19) should be made for the case of large $R_H R_\sigma$.

The Squire's theorem of the stability of ordinary hydrodynamics does not hold here, i. e., for equations (7.16) to (7.20). However under the approximations of equations (7.29), the Squire's theorem then holds here too.

5. Stability of boundary layer flow. Strictly speaking, the analysis of stability problems of the last three sections holds for two-dimensional parallel basic flow only. The problem then would be very restrictive. However, it has been found that the stability theory of parallel flows is applicable to the flow which is essentially parallel to one direction, such as boundary layer flows. Thus the stability of flows of many practically important problems may be analyzed.

For most practical problems, the magnetic Reynolds number is small, hence we may use the stability equations discussed in sections 3 and 4 to study the stability of boundary layer flow. For the cases of coplanar magnetic field, the stability equation is equation (7.24) and the basic flow is exactly the same of ordinary fluid dynamics. For the cases of transverse magnetic field, the stability equation (7.29) and the basic flow velocity distribution depends on the strength of the applied magnetic field.

The calculation of stability of boundary layer flow over a semi-infinite flat plate has been carried out for both the coplanar and the transverse magnetic field cases by ROSSOW in reference *14*. In the case of coplanar magnetic field, the magnetic field has a stabilizing effect on the flow as in the case of flow in a channel but the increase in the critical Reynolds number is small in comparison with its increase in the case of flow in channel for the same quantity $R_H R_\sigma$.

For the case of transverse magnetic field, ROSSOW studied two different cases: one is a transverse magnetic field fixed relative to the flat plate and the other is a transverse magnetic field fixed relative to the fluid far away from the plate. For the case of magnetic field fixed relative to the plate, the magnetic field has a destablizing effect because of the velocity profile having an inherently unstable shape. For the case of magnetic field fixed relative to the fluid far away from the plate, the magnetic field has a stabilizing effect. The conclusions about the difference of these two cases are doubtful because of the neglect of induced electromagnetic fields [cf. reference *15* of chapter VI].

The stability of a laminar jet mixing region under coplanar magnetic field has been studied by CURLE in reference *5*. He also found that the magnetic field has a stabilizing effect. Complete stability is predicted for large values of $R_H R_\sigma$. Of course this result is also subjected to the question of the validity of the analysis for large values of $R_H R_\sigma$.

The stability analysis for arbitrary $R_H R_\sigma$ in which the induced magnetic field is not negligible has not been studied. It should be pointed out that the basic flow over a semi-infinite plate for arbitrary $R_H R_\sigma$ in which the induced magnetic field is not negligible has also not been studied.

6. Stability of laminar flow between rotating cylinders (*2*). The most successful investigation of stability problem is probably Taylor's treatment of the Couette flow between concentric cylinders for an incompressible fluid in which the centrifugal force plays a dominant part. CHANDRASEKHAR (*2*) has extended Taylor's case for electrically conducting fluid under the influence of an external applied axial magnetic field and EDMONDS (*6*) has extended Taylor's case for an electrically conducting fluid subjected to circular magnetic field. We are going to discuss these problems here.

In the present problem, it is convenient to write our fundamental equations (4.47) and (4.9) in cylindrical coordinates (r, θ, z) with the radial, the tangential and the axial velocity component denoted by u_r, u_θ, u_z and the corresponding components of magnetic field by H_r, H_θ, H_z respectively. Our fundamental equations are as follows:

$$\frac{D u_r}{D t} - \frac{u_\theta^2}{r} - \frac{\mu_e}{\rho}\left\{(\vec{H}\cdot\nabla) H_r - \frac{H_\theta^2}{r}\right\} = -\frac{\partial p_t}{\partial r} + \nu\left(\nabla^2 u_r - \frac{2}{r^2}\cdot\frac{\partial u_\theta}{\partial \theta} - \frac{u_r}{r^2}\right)$$

$$\frac{D u_\theta}{D t} + \frac{u_\theta u_r}{r} - \frac{\mu_e}{\rho}\left\{(\vec{H}\cdot\nabla) H_\theta + \frac{H_\theta H_r}{r}\right\} = -\frac{1}{r}\cdot\frac{\partial p_t}{\partial \theta} + \nu\left(\nabla^2 u_\theta + \frac{2}{r^2}\cdot\frac{\partial u_r}{\partial \theta} - \frac{u_\theta}{r^2}\right)$$

$$\frac{D u_z}{D t} - \frac{\mu_e}{\rho}(\vec{H}\cdot\nabla) H_z = -\frac{\partial p_t}{\partial z} + \nu\nabla^2 u_z$$

$$\frac{D H_r}{D t} - (\vec{H}\cdot\nabla) u_r = \nu_H\left(\nabla^2 H_r - \frac{2}{r^2}\cdot\frac{\partial H_\theta}{\partial \theta} - \frac{H_r}{r^2}\right) \tag{7.30}$$

$$\frac{D H_\theta}{D t} - (\vec{H}\cdot\nabla) u_\theta + \frac{1}{r}(u_\theta H_r - u_r H_\theta) = \nu_H\left(\nabla^2 H_\theta + \frac{2}{r^2}\cdot\frac{\partial H_r}{\partial \theta} - \frac{H}{r^2}\right)$$

$$\frac{D H_z}{D t} - (\vec{H}\cdot\nabla) u_z = \nu_H \nabla^2 H_z$$

$$\frac{\partial u_r}{\partial r} + \frac{u_r}{r} + \frac{1}{r}\cdot\frac{\partial u_\theta}{\partial \theta} + \frac{\partial u_z}{\partial z} = 0$$

$$\frac{\partial H_r}{\partial r} + \frac{H_r}{r} + \frac{1}{r}\cdot\frac{\partial H_\theta}{\partial \theta} + \frac{\partial H_z}{\partial z} = 0$$

where $\frac{D}{Dt} = \frac{\partial}{\partial t} + (\vec{u}\cdot\nabla) = \frac{\partial}{\partial t} + u_r\frac{\partial}{\partial r} + \frac{u_\theta}{r}\cdot\frac{\partial}{\partial \theta} + u_z\frac{\partial}{\partial z}$

$$\vec{H}\cdot\nabla = H_r\frac{\partial}{\partial r} + \frac{H_\theta}{r}\cdot\frac{\partial}{\partial \theta} + H_z\frac{\partial}{\partial z}$$

$$\nabla^2 = \frac{\partial}{\partial r^2} + \frac{1}{r}\cdot\frac{\partial}{\partial r} + \frac{1}{r^2}\cdot\frac{\partial^2}{\partial \theta^2} + \frac{\partial^2}{\partial z^2}$$

$$p_t = \frac{p}{\rho} + \mu_e\frac{\vec{H}\cdot\vec{H}}{\rho}$$

(a) The case of external applied axial magnetic field. In this case, the basic flow has the following values:

$$u_r = u_z = 0, \quad u_\theta = V(r) = A r + \frac{B}{r}, \quad \frac{\partial p_t}{\partial r} = \frac{V^2}{r}$$
$$H_r = H_\theta = 0, \quad H_z = H_0 = \text{constant} \tag{7.31}$$

$$\text{where } A = \Omega_1 \frac{1 - \dfrac{m R_2^2}{R_1^2}}{1 - \dfrac{R_2^2}{R_1^2}}, \quad B = \Omega_1 \frac{R_1^2(1-m)}{1 - \dfrac{R_1^2}{R_2^2}}, \quad m = \frac{\Omega_2}{\Omega_1} \tag{7.32}$$

Ω_1 and Ω_2 are the angular velocities of the outer and the inner cylinders respectively; R_1 and R_2 are their radii respectively.

In the disturbed motion, we have

$$u_r,\ V(r)+u_\theta,\ u_z,\ h_r,\ h_\theta,\ H+h_z,\ \frac{V^2}{r}+p_t' \tag{7.33}$$

where u_r, u_θ, u_z, h_r, h_θ, h_z and p_t' are small quantities. We assume that the perturbed velocity and magnetic field are of the following form:

$$\begin{aligned} u_r &= e^{\omega t}\, u(r) \cos\lambda z, & h_r &= e^{\omega t}\, h_1(r) \sin\lambda z \\ u_\theta &= e^{\omega t}\, v(r) \cos\lambda z, & h_\theta &= e^{\omega t}\, h_2(r) \sin\lambda z \\ u_z &= e^{\omega t}\, w(r) \cos\lambda z, & h_z &= e^{\omega t}\, h_3(r) \sin\lambda z \end{aligned} \tag{7.34}$$

Substituting equations (7.33) and (7.34) into equation (7.30) and neglecting the high order terms, we have the linearized equations for the disturbances: In these equations, it is possible to eliminate the variables ω, h_3 and p_t' and finally we have the following system of stability equations for our analysis:

$$\begin{gathered} \frac{\nu}{\lambda^2}\left(D D_1-\lambda^2-\frac{\omega}{\nu}\right) v-2 A u+\frac{\mu_e H \lambda}{\rho} h_2=0 \\ \lambda H u=\nu_H\left(D D_1-\lambda^2-\frac{\omega}{\nu}\right) h_1 \\ \lambda H v=\nu_H\left(D D_1-\lambda^2-\frac{\omega}{\eta}\right) h_2-\frac{2 B}{r^2} h_1 \\ \frac{\nu}{\lambda^2}\left(D D_1-\lambda^2-\frac{\omega}{\nu}\right) \cdot\left(D D_1-\lambda^2\right) u+\frac{\mu_e H}{\rho \lambda}\left(D D_1-\lambda^2\right) h_1=2\left(A+\frac{B}{r^2}\right) v \end{gathered} \tag{7.35}$$

where $D = \mathrm{d}/\mathrm{d}r$ and $D_1 = D + 1/r$.

The boundary conditions are

$$u = v = D_1 u = h_1 = D_1 h_2 = 0 \quad \text{at} \quad r = R_1 \quad \text{and} \quad r = R_2 \tag{7.36}$$

where the boundary conditions for the magnetic field are known as Fermi boundary conditions in which the cylinders are supposed to be perfect electrical conductors so that the z-component of electric field and the r-component of the magnetic field vanish on the surface of the cylinders.

In general the factor ω is complex, i. e., $\omega = \omega_R + i\,\omega_i$. The condition in marginal stability is $\omega_R = 0$. If ω_i is also equal to zero, we call this case the principle of exchange of stability if $\omega_i \neq 0$, we call it over-stability. The equations governing marginal stability with the principle of exchange of stabilities are then equations (7.35) with $\omega = 0$.

The general solution of the system of equations (7.35) with $\omega = 0$ is very difficult. Chandrasekhar considered the following case with proper approximations:

(a) The gap between the cylinders is small compared with the mean radius, i. e.,

$$\mathrm{d} = (R_2 - R_1) \ll \tfrac{1}{2}(R_2 + R_1) = R_0 \tag{7.37}$$

Under this condition, we need not distinguish between D and D_1.

(b) The two cylinders are rotating in the same direction. We can make the following approximations:

$$A+\frac{B}{r}\cong A+\frac{B}{R_0}=\tfrac{1}{2}\Omega_1\left[(1+m)-\frac{3\,\mathrm{d}}{4R_0}(1-m)\right]=\Omega_0$$
$$\frac{B}{r^2}\cong\frac{B}{R_0{}^2} \tag{7.38}$$

(c) ν_H is much larger than ν. The third equation of (7.35) becomes

$$v=\frac{\nu_H}{H\,a\,\mathrm{d}}(D^2-a^2)\,h_2 \tag{7.39}$$

Under these conditions, equation (7.35) may be reduced to a single differential equation for h_2 of order eight, as follows:

$$[(D^2-a^2)^2+Q\,a^2]^2\,h_2=-T\,a^2\,(D^2-a^2)\,h_2 \tag{7.40}$$

$$\text{where } D=\frac{\mathrm{d}}{\mathrm{d}\zeta},\quad \zeta=\frac{1}{\mathrm{d}}(r-R_0),\quad \lambda=\frac{a}{\mathrm{d}}$$

$$Q=\frac{\mu_e{}^2\,H_0{}^2\,\sigma}{\rho\,\nu}\,\mathrm{d}^2 \tag{7.41}$$

$$T=-\frac{4\,\Omega_0\,A}{\nu^2}\,\mathrm{d}^4=\text{Taylor number} \tag{7.42}$$

The corresponding boundary conditions are

$$\text{at } \zeta=\pm\tfrac{1}{2}:$$
$$[(D^2-a^2)^2+Q\,a^2]\,h_2=0,\quad D\,[(D^2-a^2)^2+Q\,a^2]\,h_2=0 \tag{7.43}$$
$$(D^2-a^2)\,h_2=0,\quad D\,h_2=0$$

Chandrasekhar (*1*) solved the eigenvalue problem of equation (7.40) with the boundary conditions (7.43) by the variational method and also by the method of an expansion in orthogonal functions. He found that the magnetic field has a stabilizing effect. As Q increases, the critical value of the Taylor number increases.

(b) The case of a circular magnetic field. Edmonds considered the stability of flow between rotating cylinders in a circular magnetic field. The basic flow has the following values:

$$u_r=u_z=0,\quad u_\theta=V(r)=A\,r+\frac{B}{r},\quad \frac{\partial p_t}{\partial r}=\frac{V^2}{r}-\frac{\mu_e}{\rho}\cdot\frac{H^2}{r}$$
$$H_r=H_z=0,\quad H_\theta=H(r)=C\,r+\frac{G}{r} \tag{7.44}$$

where A, B, C and G are arbitrary constants to be determined by boundary conditions.

Edmonds (*6*) used essentially the same method to discuss the stability of this flow. He also found that the magnetic field inhibits the onset of instability but this effect is quite small as the hydromagnetic interaction involves displacement but not distortion of the magnetic lines of force.

7. Stability of a layer of fluid heated below (*1*, *3*, *4*). This problem is particularly interesting in many astrophysical and geophysical problems (*4*). It was first investigated by Rayleigh, Jeffreys and others for ordinary fluid dynamics. The main results are (i) that the fluid becomes unstable when the Rayleigh number (7.67) exceeds a certain value and (ii) that the motion must have a cellular pattern when instability occurs. This has been verified experimentally. Chandrasekhar extended this problem to a conducting fluid in a magnetic field.

Another interesting point of this problem is that in ordinary fluid dynamics, the present problem and that of rotating cylinders are closely related from the mathematical point of view. Taylor's criterion (7.42) for the onset of rotational instability in the case when the two cylinders are rotating in the same direction can be deduced from Rayleigh's criterion for thermal instability by replacing $\varkappa \nu$ by ν^2. In magnetohydrodynamics, the same close analogy between these two problems does not exist but in both cases the effect of the magnetic field inhibits the onset of instability.

We are going to discuss this problem according to Chandrasekhar's analysis (*3*).

The equation of motion is essentially equation (4.47b) except for the following two points:

(i) The gravitational force $\vec{F}_g$ should be added to equation (4.47b), i. e.,

$$\vec{F}_g = -\rho g \vec{\lambda} \tag{7.45}$$

where $\vec{\lambda}$ is a unit vector in the vertical direction.

(ii) The density in equation (7.45) may be written as

$$\rho = \rho_0 (1 - \alpha \Delta T) \tag{7.46}$$

where ρ_0 is the average density which corresponds to a temperature T_0. In the acceleration term $\rho\, D u/D t$, we may still put $\rho = \rho_0$. Here α is the coefficient of volume expansion. The local temperature deviation is

$$\Delta T = T - T_0 \tag{7.47}$$

The local temperature may be written as

$$T = T_0 + \beta \lambda_j x_j + \theta \tag{7.48}$$

where a constant mean temperature gradient $\beta = -|\beta|$ is assumed to be maintained by an external agency.

$$\beta = \frac{\partial \overline{T}}{\partial x_j} \lambda_j \tag{7.49}$$

θ is the deviation of temperature from the local mean value $T_0 + \beta \lambda_j x_j$.

The equation of motion for our problem is then

$$\frac{\partial u_i}{\partial t} + \frac{\partial u_i u_j}{\partial x_j} - \frac{\mu_e}{\rho_0} H_j \frac{\partial H_i}{\partial x_j} = \nu \nabla^2 u_i + \gamma \theta \lambda_i - \frac{\partial p_t}{\partial x_i} \tag{7.50}$$

$$\text{where } \gamma = \alpha g \tag{7.51}$$

$$\text{and } p_t = \frac{p}{\rho_0} + \frac{\mu_e H^2}{2 \rho_0} + g \lambda_j x_j - \tfrac{1}{2} \beta \gamma \lambda_i \lambda_j x_i x_j \tag{7.52}$$

The energy equation (4.49a) is

$$\frac{\partial \theta}{\partial t} + \beta \lambda_j u_j + u_j \frac{\partial \theta}{\partial x_j} = \varkappa_1 \nabla^2 \theta \tag{7.53}$$

where $\varkappa_1 = \varkappa \rho_0 / C_v$. Both $\varkappa$ and C_v are assumed to be constant here.

The equation for the magnetic field is still equation (4.47c). Now we assume that the external magnetic field is a constant in the direction of the unit vector $\vec{\tau}$. The total magnetic field is then

$$H_i = H \tau_i + h_i \tag{7.54}$$

where H is the magnitude of the external magnetic field and $\vec{h}$ is the induced magnetic field due to the motion of the plasma. The equation for the magnetic field is then

$$\frac{\partial h_i}{\partial t} + u_j \frac{\partial h_i}{\partial x_j} - (H \tau_j + h_j) \frac{\partial u_i}{\partial x_j} = \nu_H \nabla^2 h_i \tag{7.55}$$

We assume that θ, u_i and h_i are small quantities. Thus we may neglect the high order terms in equations (7.50), (7.53) and (7.55) in order to obtain the stability equations for θ, u_i and h_i. CHANDRASEKHAR found that it is convenient to introduce the new variables:

$$w = u_i \lambda_i, \quad h = h_i \lambda_i \tag{7.56}$$

The linearized equation for the variables, w and h are then

$$\frac{\partial \theta}{\partial t} = - \beta w + \varkappa_1 \nabla^2 \theta \tag{7.57a}$$

$$\frac{\partial w}{\partial t} = \frac{\mu_e H}{\rho_0} \tau_j \frac{\partial h}{\partial x_j} + \gamma \theta + \nu \nabla^2 w - \lambda_i \frac{\partial p_t}{\partial x_i} \tag{7.57b}$$

$$\frac{\partial h}{\partial t} = H \tau_j \frac{\partial w}{\partial x_j} + \nu_H \nabla^2 h \tag{7.57c}$$

If we consider a horizontal layer of fluid of depth d, confined between two parallel planes at $z = 0$ and $z = d$. The boundary conditions are

$$\theta = w = 0 \quad \text{at} \quad z = 0 \quad \text{and} \quad z = d$$

$$\frac{\partial^2 w}{\partial z^2} = 0 \text{ on a free surface} \tag{7.58}$$

$$\frac{\partial w}{\partial z} = 0 \text{ on a rigid surface}$$

According to the principle of the exchange of stabilities, we obtain the following equation for marginal stability from equation (7.57):

$$\nabla^6 w = \frac{|\beta| \gamma}{\varkappa_1 \nu} \left(\nabla^2 w - \lambda_i \lambda_j \frac{\partial^2 w}{\partial x_i \partial x_j} \right) + \frac{\mu_e^2 H^2 \sigma}{\rho_0 \nu} \tau_i \tau_j \frac{\partial^2 \nabla^2 w}{\partial x_i \partial x_j} \tag{7.59}$$

The equation for θ is of exactly the same form as w.

If both the external magnetic field and the gravitational force are in the z-direction, i. e., $\vec{\tau} = \vec{\lambda} = (0, 0, 1)$, equation (7.59) becomes

$$\nabla^6 w = \frac{|\beta| \gamma}{\varkappa \nu} \left(\frac{\partial^2 w}{\partial x^2} + \frac{\partial^2 w}{\partial y^2} \right) + \frac{\mu_e^2 H^2 \sigma}{\rho_0 \nu} \cdot \frac{\partial^2 \nabla^2 w}{\partial z^2} \tag{7.60}$$

We may use the method of separation of variables to solve equation (7.60). We write

$$w = f(x, y) \, W(z) \tag{7.61}$$

$$\text{where } d^2 \left(\frac{\partial^2 f}{\partial x^2} + \frac{\partial^2 f}{\partial y^2} \right) = -a^2 f \tag{7.62}$$

where a^2 is a certain "characteristic number" determined by the conditions on the cell walls. The solution for f when the boundary of the cell is a rectangle of sides L_1 and L_2 is

$$f = A_{mn} \cos \frac{m \pi x}{L_1} \cos \frac{n \pi y}{L_2} \tag{7.63}$$

where m and n are arbitrary even integers and A_{mn} is a constant, and

$$a^2 = d^2 \pi^2 \left(\frac{m^2}{L_1} + \frac{n^2}{L_2} \right) \tag{7.64}$$

For other shapes of cell, different solutions may be obtained.

The equation for $W(z) = W(\zeta/d)$ is

$$(D^2 - a^2) \cdot [(D^2 - a^2)^2 - Q D^2] \, W = - a^2 R W \tag{7.65}$$

where $D = d/d\zeta$ and

$$Q = \frac{\mu_e H^2 \sigma}{\rho_0 \nu} d^2 \tag{7.66}$$

$$R = \frac{|\beta| \gamma}{\varkappa \nu} d^4 = \text{Rayleigh number} \tag{7.67}$$

It is interesting to notice the similarity in form of the Rayleigh number (7.67) and that of the Taylor number (7.42). For the effect of magnetic field, the same factor Q holds for both cases [cf. (7.41)].

The six boundary conditions for W are:

$$\begin{aligned} &\text{at } \zeta = 0 \text{ and } \zeta = 1 \\ &W = [(D^2 - a^2)^2 - Q\, D^2]\, W = 0 \\ &D\, W = 0 \text{ on a rigid surface} \\ &D^2\, W = 0 \text{ on a free surface} \end{aligned} \tag{7.68}$$

In principle, we may solve equation (7.65) with the boundary conditions (7.68) but a tremendous amount of work is involved. CHANDRASEKHAR used a variational principle to determine the critical Rayleigh numbers for various cases in reference *1*. He found that under normal terrestial conditions ($\nu_H > \varkappa_1$) the inhibition of convection by a magnetic field is a very pronounced effect.

8. Stability of pinch effect. In controlled fusion research, one of the most interesting problems is to confine the plasma by electromagnetic field (*13*, *15*).

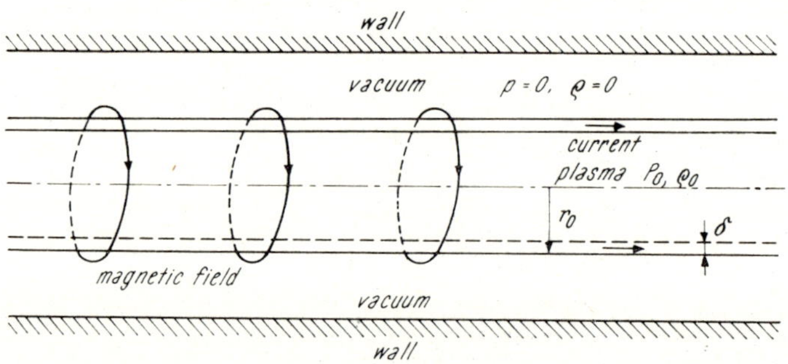

Fig. 7.1. Ideal Pinch Effect

A simple example of such a confinement is the classical pinch effect. Fig. 7.1 shows a sketch of the ideal pinch effect. The plasma lies within an infinitely long circular cylinder of radius r_0. There is a uniform sheet current of thickness δ on the cylinder parallel to the axis, which produces a toroidal magnetic field outside the cylinder. The plasma pressure is balanced by the magnetic pressure. In the separating surface of width δ, the current density $\vec{J}$ and the excess electrical change ρ_e are finite but large of order δ^{-1} and across this surface the other physical quantities vary continuously. The integrals of $\vec{J}$ and ρ_e across this surface are given by $\vec{J}^*$ and $\rho_e{}^*$ respectively.

A detailed analysis of this development of a self-pinched plasma has not been carried out. In this section we shall show a simple analysis of the stability of an ideal pinch effect of Fig. 7.1 as first given by KRUSKAL and SCHWARZSCHILD in reference *7*. They consider an ideal plasma with $\sigma = \infty$, $\mu = \varkappa = 0$, but they did not neglect the displacement current $\partial\, \varepsilon\, \vec{E}/(\partial\, t)$ nor the excess electrical charge ρ_e. The variables in this problem are then $\vec{E}$, $\vec{B} = \mu_e\, \vec{H}$, $\vec{J}$, ρ_e, $\vec{u}$, p and ρ. The fundamental equations are then

$$\rho \frac{D\vec{u}}{D t} = \vec{J} \times \vec{B} + \rho_e\, \vec{E} - \nabla\, p \tag{7.69a}$$

$$\nabla \cdot (\rho \vec{u}) = -\frac{\partial \rho}{\partial t} \tag{7.69b}$$

$$\vec{E} + \vec{u} \times \vec{B} = \frac{1}{\sigma}(\vec{J} - \rho_e \vec{u}) \tag{7.69c}$$

$$\nabla \times \vec{B} = \mu_e \left(\vec{J} + \varepsilon \frac{\partial \vec{E}}{\partial t}\right) \tag{7.69d}$$

$$\nabla \cdot \vec{B} = 0 \tag{7.69e}$$

$$\nabla \times \vec{E} = -\frac{\partial \vec{B}}{\partial t} \tag{7.69f}$$

$$\nabla \cdot \vec{E} = \frac{1}{\varepsilon} \rho_e \tag{7.69g}$$

$$\frac{1}{p} \cdot \frac{D p}{D t} = \frac{\gamma}{\rho} \cdot \frac{D \rho}{D t} \tag{7.69h}$$

The basic flow conditions are as follows:
(a) Within the cylinder $r < r_0$, we have

$$p = p_0, \quad \rho = \rho_0, \quad \vec{u} = \vec{E} = \vec{B} = \vec{J} = \rho_e = 0 \tag{7.70a}$$

(b) In the vacuum $r > r_0$

$$p = \rho = \vec{u} = \vec{J} = \vec{E} = B_r = B_z = 0, \quad B_\theta = B_0 \, r_0/r \tag{7.70b}$$

where B_r, B_z and B_θ are the r-, z- and θ-component of $\vec{B}$ respectively.
(c) On the surface $r = r_0$

$$J_r^* = J_\theta^* = \rho_e = 0, \quad J_z^* = J_0^* \tag{7.70c}$$

From the equilibrium condition of the plasma pressure and the magnetic pressure on the surface $r = r_0$, we have

$$B_0 = \mu_e J_0^*, \quad J_0^* B_0 = 2 \, p_0 \tag{7.70d}$$

Now we consider small perturbances from the equilibrium by a tilde. The perturbed quantities q' have the following form:

$$q' = q_1(r) \exp(i\,\theta + i\,k\,z + \omega\,t) \tag{7.71}$$

where q' may be p', B_z', etc.

Substituting the expressions of the perturbed motion into the fundamental equation (7.69) and neglecting the high order terms of the perturbed quantities q_1, we have the linearized equations for the disturbances. After eliminations, we have in the plasma

$$r \frac{d}{dr}\left(r \frac{d}{dr} p_1\right) - \left[\left(k^2 + \frac{\rho_0 \, \omega^2}{\gamma \, p_0}\right) r^2 + 1\right] p_1 = 0 \tag{7.72}$$

and in the vacuum

$$r \frac{d}{d r}\left(r \frac{d}{d r} B_{z1}\right) - [(k^2 + \mu_e \varepsilon \omega^2) r^2 + 1] B_{z1} = 0 \tag{7.73}$$

The solutions of equations (7.72) and (7.73) are

$$p_1 = C_p J(\zeta r), \quad B_{z1} = C_B H(\eta r) \tag{7.74}$$

where C_p and C_B are constants and

$$\zeta^2 = k^2 + \frac{\rho_0 \omega^2}{\gamma p_0}, \quad \eta^2 = k^2 + \mu_e \varepsilon \omega^2 \tag{7.75}$$

$$J(x) = - i J_1(i x), \quad H(x) = - H_1^{(1)}(i x) \tag{7.76}$$

J_1 and $H_1^{(1)}$ are the Bessel and Hankel functions of the first kind and of first order. Here we assume that ω^2 is real and take $\zeta \geqq 0$, $\eta \geqq 0$.

The other quantities E_{z1}, ρ', etc. may be easily obtained.

Kruskal and Schwarzschild found that on the separating surface the following relation holds

$$\frac{\rho_0 r_0^2 \eta \omega^2}{2 p_0 \rho} \cdot \frac{J(\zeta r_0)}{J'(\zeta r_0)} = \eta r_0 + \frac{k^2}{\eta^2} \cdot \frac{H(\eta r)}{H'(\eta r_0)} + \mu_e \varepsilon r_0^2 \omega^2 \frac{H'(\eta r)}{H(\eta r_0)} \tag{7.77}$$

If we neglect terms with $1/c^2 = \mu_e \varepsilon$, equation (7.77) becomes with $\eta = |k|$

$$\frac{\rho_0 r_0^2 |k| \omega^2}{2 p_0 \zeta} \cdot \frac{J(\zeta r_0)}{J'(\zeta r_0)} = |k| r_0 + \frac{H(|k| r_0)}{H'(|k| r_0)} \tag{7.78}$$

For any fixed value of $k \neq 0$, the left-hand side of equation (7.78) varies monotonically from zero to infinity as ω^2 goes from 0 to ∞ through real values; since the right hand side can be shown to be positive, it follows that equation (7.78) has a unique real positive solution for ω^2 for each value of k, thus the equilibrium is unstable.

To stabilize the pinch effect under various conditions is one of the most interesting current research problem (*15*).

9. Experiment of stability of flow in magnetohydrodynamics. Lehnert (*8*) reported some experimental results of stability of flow in a magnetic field. In his experiment, he found that a shallow layer of mercury over a copper disk with two concentric copper rings is destabilized by applying a vertial magnetic field. The rotation of the inner copper ring produces a shear layer in the mercury which is intensified by the magnetic field to the extent that eddy type flow results. Even though this experiment shows that the magnetic field has a destabilizing effect, it does not mean in general that the magnetic field has a destabilizing effect. As we have shown in the last few sections, the magnetic field has a stabilizing effect in certain cases. Hence as Lehnert pointed out, the effect of the magnetic field on the stability of flow for each special case whould be studied individually in order to determine whether the magnetic field has a stability or destabilizing effect.

References

1. CHANDRASEKHAR, S.: On the Inhibition of Convection of a Magnetic Field. Phil. Mag. ser. 7, vol. 43, No. 340, May 1952, pp. 501–532.
2. CHANDRASEKHAR, S.: The Stability of Viscous Flow between Rotating Cylinders to the Presence of a Magnetic Field. Proc. Roy Soc. London A-216, 10 Feb. 1953, pp. 293–309.
3. CHANDRASEKHAR, S.: The Instability of a Layer of Fluid Heated below and Subject to the Simultaneous Action of a Magnetic Field and Rotation. Proc. Roy. Soc. London A-225, no. 1161, 31 Aug. 1954, pp. 173–184.
4. COWLING, T. G.: Magnetohydrodynamics. Chap. IV. Interscience Publishers, Inc., N.Y., 1957.
5. CURLE, N.: On the Stability in the Presence of a Coplanar Magnetic Field of a Laminar Mixing Region in a Viscous Electrically Conducting Fluid. British ARC Rep. 18720, FM 2461, 30 Oct. 1956.
6. EDMONDS, F. N. Jr.: Hydromagnetic Stability of a Conducting Fluid in a Circular Magnetic Field. The Physics of Fluids, vol. 1, No. 1, Jan.–Feb. 1958.
7. KRUSAL, M., and M. SCHWARZSCHILD: Some Instabilities of a Completely Ionized Plasma. Proc. Roy. Soc. London A-223, 1954, pp. 348–360.
8. LEHNERT, B.: An Instability of Laminar Flow of Mercury Caused by an External Magnetic Field. Proc. Roy. Soc. London A-233, 29 Dec. 1955, pp. 299–302.
9. LIN, C. C.: The Theory of Hydrodynamic Stability. Cambridge Univ. Press, 1955.
10. LOCK, R. C.: The Stability of the Slof of an Electrically Conducting Fluid between Parallel Planes under a Transverse Magnetic Field. Proc. Roy. Soc. London A-233, Dec. 1955, pp. 105–125.
11. MICHAEL, D. H.: Stability of Plane Parallel Flows of Electrically Conducting Fluids. Proc. Cambridge Phil. Soc., vol. 49, pt. 1, Jan. 1953, pp. 166–168.
12. PAI, S. I.: Viscous Flow Theory. I–Laminar Flow, Chap. XII. D. Van Nostrant Co., Inc., Princeton, N.Y., 1956.
13. POST, R. F.: Controlled Fusion Research-an Application of the Physics of High Temperature Plasmas. Rev. of Mod. Phys., vol. 28, No. 3, July 1956, pp. 338–362.
14. ROSSOW, V. J.: Boundary Layer Stability Diagrams for Electrically Conducting Fluids in the Presence of a Magnetic Field. N. A. C. A. Tech. Note No. 4282, 1958.
15. SPITZER, L. Jr.: Physics of Fully Ionized Gases. Interscience Publishers, Inc., N.Y., 1956.
16. STUART, J. T.: On the Stability of Viscous Flow between Parallel Planes in the Presence of a Coplanar Magnetic Field. Proc. Roy. Soc. London A-221, Jan. 1954, pp. 189–206.

Chapter VIII

Turbulence in Magnetohydrodynamics

1. Introduction. The most interesting and probably the least understood flow problem is the turbulent flow. The essential characteristic of turbulent flow is that the turbulent fluctuations are random in nature; hence, the final and logical solution of the turbulence problem requires the application of the methods of statistical mechanics. Considerable progress has been made in this respect during the last twenty years but there are still many important problems to be solved before a successful statistical theory of turbulence can be developed (*7*). At the present time, most of the results from the statistical theory of turbulence are concerned with isotropic turbulence which was first investigated by G. I. TAYLOR and which was further developed by VON KÁRMÁN, KOLMOGOROFF, HEISENBERG and others. In magnetohydrodynamics, the problem is much more complicated, because we have to consider both the turbulent fluctuations in the magnetic field as well as those in the velocity field and their interaction. It was CHANDRASEKHAR who extended the ordinary statistical theory of isotropic turbulence to the case of magnetohydrodynamics (*2*). He was able to show that many well-known results of the invariant theory of isotropic turbulence can be extended in magnetohydrodynamics. We are going to discuss Chandrasekhar's statistical theory of isotropic turbulence in this chapter (§§ 2 to 6, and 8).

The interaction between a magnetic field and turbulent motion has frequently been studied in connection with the possible existence of an interstellar magnetic field. In a medium of high electrical conductivity, turbulence will be accompanied by a spontaneous generation and a subsequent amplification of a magnetic field. BATCHELOR (*1*) discussed in detail the behavior of a spontaneous magnetic field in turbulent flow without an external magnetic field which will be briefly discussed in § 7. In a state of equilibrium, the energy per unit volume in the magnetic and the velocity fields will approach equality. Chandrasekhar's theory (*4*) of the partition of energy in hydromagnetic turbulence will be discussed in § 9.

2. Double and triple correlation in isotropic turbulence. In Taylor-Kármán statistical theory (*7*), the important notion is that of studying the correlation between two fluctuating quantities in turbulent flow. We take a system of rectangular Cartesian coordinates x_i $(i = 1, 2, 3)$ and denote the turbulent velocity components by u_i. Let $\vec{A}$ and $\vec{B}$ be two fixed vectors with components A_i and B_i respectively; P and P' be the points with coordinates x_i and x_i' in a field of turbulent motion respectively; the vector distance $\overline{pp'} = \vec{r}$, and the components of $\vec{r}$ be ζ_i so that $\zeta_i = x_i' - x_i$.

The correlation between the turbulent velocity component u_A in the direction of the vector $\vec{A}$ at the point P and the turbulent velocity component u_B' in the direction of the vector $\vec{B}$ at point P' is

$$\overline{u_A\, u_B'} = \frac{A_i\, B_j}{|A|\,|B|}\, \overline{u_i\, u_j'} = \frac{A_i\, B_j}{|A|\,|B|}\, R_{ij} \tag{8.1}$$

where the double suffix summation convection is employed. This correlation is therefore a second order tensor, with components $R_{ij} = \overline{u_i\, u_j'}$.

We may also define triple correlation of velocity components as follows:

$$T_{ijk} = \overline{u_i\, u_j\, u_k'} \tag{8.2}$$

This triple correlation T_{ijk} is a tensor of third order. $i, j, k = 1, 2,$ or 3.

In magnetohydrodynamics, we have to consider turbulent fluctuations of a magnetic field. CHANDRASEKHAR introduced the new variable $\vec{h}$

$$\vec{h} = \left(\frac{\mu_e}{\rho}\right)^{\frac{1}{2}} \vec{H} \tag{8.3}$$

for the magnetic field $\vec{H}$. Hence $\vec{h}$ is essentially the velocity of Alfvén's wave. For magnetohydrodynamics, instead of correlations of velocity components only we have also correlation between velocity components and h_i which characterizes the magnetic field. Thus in magnetohydrodynamics, we have the double correlations

$$\overline{u_i\, u_j'}, \quad \overline{h_i\, h_j'} \text{ and } \overline{u_i\, h_j'}$$

and the triple correlations

$$\overline{u_i\, u_j\, u_k'}, \quad \overline{h_i\, h_j\, h_k'}, \quad \overline{u_i\, u_j\, h_k'}, \quad \overline{h_i\, h_j\, u_k'}$$

$$\overline{(h_i\, u_j - u_i\, h_j)\, h_k'} \text{ and } \overline{u_i\, (h_j'\, u_k' - h_k'\, u_j')}$$

In isotropic turbulence, each of these double and triple correlations depends on one scalar function because the divergence of both $\vec{u}$ and $\vec{h}$ are zero. These scalar functions are functions of r and t. In Chandrasekhar's notation we have the following scalar functions:

$$\overline{u_i\, u_j'} = \frac{Q'}{r}\, \zeta_i\, \zeta_j - (r\, Q' + 2\, Q)\, \delta_{ij} \tag{8.4a}$$

$$\overline{h_i\, h_j'} = \frac{H'}{r}\, \zeta_i\, \zeta_j - (r\, H' + 2\, H)\, \delta_{ij} \tag{8.4b}$$

$$\overline{u_i\, h_j'} = R\, \varepsilon_{ijl}\, \zeta_l \tag{8.4c}$$

$$\overline{u_i\, u_j\, u_k'} = \frac{2}{r}\, T'\, \zeta_i\, \zeta_j\, \zeta_k - (r\, T' + 3\, T) \cdot (\zeta_i\, \delta_{jk} + \zeta_j\, \delta_{ik}) + 2\, T\, \delta_{ij}\, \zeta_k \tag{8.4d}$$

$$\overline{h_i\, h_j\, u_k'} = \frac{2}{r}\, S'\, \zeta_i\, \zeta_j\, \zeta_k - (r\, S' + 3\, S) \cdot (\zeta_i\, \delta_{jk} + \zeta_j\, \delta_{ik}) + 2\, S\, \delta_{ij}\, \zeta_k \tag{8.4e}$$

$$\overline{u_i\, u_j\, h_k'} = U\,(\zeta_i\, \varepsilon_{jkl}\, \zeta_l + \zeta_j\, \varepsilon_{ikl}\, \zeta_l) \tag{8.4f}$$

$$\overline{h_i\, h_j\, h_k'} = V\,(\zeta_i\, \varepsilon_{jkl}\, \zeta_l + \zeta_j\, \varepsilon_{ikl}\zeta_l) \tag{8.4g}$$

$$\overline{(h_i\, u_j - h_j\, u_i)\, h_k'} = P\,(\zeta_i\, \delta_{jk} - \zeta_j\, \delta_{ik}) \tag{8.4h}$$

$$\overline{u_i\,(h_j'\, u_k' - h_k'\, u_j')} = (2\,W + r\,W')\, \varepsilon_{ijk} - \frac{W'}{r}\, \zeta_i\, \varepsilon_{jkl}\, \zeta_l \tag{8.4i}$$

where a prime on the scalar function means differentiation with respect to r, e. g., $Q' = \partial\, Q/\partial\, r$. $\delta_{ij} = 0,\ i \neq j;\ \delta_{ij} = 1,\ i = j$, etc.

3. Generalized von Kármán-Howarth equation. In ordinary statistical theory of isotropic turbulence, there is a differential equation connecting the double and the triple correlations which is known as the von Kármán-Howarth equation. In magnetohydrodynamics, similar equation has been found by CHANDRASEKHAR. Equations (4.47b) and (4.47c) in terms of u_i and h_i become

$$\frac{\partial\, u_i}{\partial\, t} + \frac{\partial}{\partial\, x_k}\,(u_i\, u_k - h_i\, h_k) = -\frac{1}{\rho} \cdot \frac{\partial}{\partial\, x_i}\,(p + \tfrac{1}{2}\, h^2) + \nu\, \nabla^2\, u_i \tag{8.5}$$

$$\frac{\partial\, h_i}{\partial\, t} + \frac{\partial}{\partial\, x_k}\,(h_i\, u_k - u_i\, h_k) = \nu_H\, \nabla^2\, h_i \tag{8.6}$$

Multiplying equation (8.5) by the velocity components u_j' at x_i' and averaging the resulting equation, we have

$$\overline{u_j' \frac{\partial\, u_i}{\partial\, t}} + \frac{\partial}{\partial\, x_k}\,(\overline{u_i\, u_k\, u_j'} - \overline{h_i\, h_k\, u_j'}) = \nu\, \nabla^2\, \overline{u_i\, u_j'} \tag{8.7}$$

The terms due to pressure correlation vanishes because of isotropy. Similarly we also have

$$\overline{u_i \frac{\partial\, u_j'}{\partial\, t}} + \frac{\partial}{\partial\, x_k'}\,(\overline{u_j'\, u_k'\, u_i} - \overline{u_j'\, u_k'\, u_i}) = \nu\, \nabla^2\, \overline{u_i\, u_j'} \tag{8.8}$$

Adding equations (8.7) and (8.8), we have

$$\frac{\partial}{\partial\, t}\,(\overline{u_i\, u_j'}) - 2\, \frac{\partial}{\partial\, x_k}\,(\overline{u_i\, u_k\, u_j'} - \overline{h_i\, u_k\, u_j'}) = 2\, \nu\, \nabla^2\, \overline{u_i\, u_j'} \tag{8.9}$$

Substituting the scalar functions defined in equation (8.4) into equation (8.9) we have the generalized von Kármán-Howarth equation

$$\frac{\partial\, Q}{\partial\, t} = 2 \left(r\, \frac{\partial}{\partial\, r} + 5 \right) \cdot (T - S) + 2\, \nu\, D\, Q \tag{8.10}$$

$$\text{where } D = \frac{\partial^2}{\partial\, r^2} + \frac{4}{r} \cdot \frac{\partial}{\partial\, r}.$$

If $S = 0$, equation (8.10) reduces to the ordinary von Kárman-Howárth equation.

From equation (8.6), we have another differential equation connecting the scalar functions H and P as follows:

$$\frac{\partial\, H}{\partial\, t} = 2\, P + 2 \cdot \nu_H\, D\, H \tag{8.11}$$

CHANDRASEKHAR also found a third differential equation connecting some of these scalar functions as follows:

$$\frac{\partial R}{\partial t} = \left(r \frac{\partial}{\partial r} + 5\right) \cdot (U - V) - D W + (\nu + \nu_H) D R \tag{8.12}$$

Of course, equations (8.11) and (8.12) occur only in magnetohydrodynamics.

4. Equations in terms of vector potential. It is sometimes convenient to use the vector potential $\vec{A}$ to replace the magnetic field $\vec{H}$ [cf. equation (3.33a)]. In terms of vector potential $\vec{A}$, we may introduce the vector $\vec{a}$ such that

$$\vec{a} = \left(\frac{1}{\mu_e \rho}\right)^{\frac{1}{2}} \vec{A} \tag{8.13}$$

The relation between $\vec{h}$ and $\vec{a}$ is then

$$\vec{h} = \nabla \times \vec{a} \tag{8.14}$$

We may define the following correlations

$$\overline{a_i a_j'}, \quad \overline{u_i a_j'}, \quad \overline{u_i u_j a_k'}, \quad \overline{h_i h_j a_k'} \quad \text{and} \quad \overline{(h_i u_j - h_j u_i) a_k'}$$

Similarly for isotropic turbulence, each of these correlations depends on one scalar function as follows:

$$\overline{a_i a_j'} = \frac{A'}{r} \zeta_i \zeta_j - (r A' + 2A) \delta_{ij} \tag{8.15a}$$

$$\overline{u_i a_j'} = B \varepsilon_{ijl} \zeta_l \tag{8.15b}$$

$$\overline{u_i u_j a_k'} = \text{curl } E (\zeta_i \varepsilon_{jkl} \zeta_l + \zeta_j \varepsilon_{ikl} \zeta_l) \tag{8.15c}$$

$$\overline{h_i h_j a_k'} = \text{curl } F (\zeta_i \varepsilon_{jkl} \zeta_l + \zeta_j \varepsilon_{ikl} \zeta_l) \tag{8.15d}$$

$$\overline{(h_i u_j - h_j u_i) a_k'} = 2 G \varepsilon_{ijk} + \frac{G'}{r} (\zeta_i \varepsilon_{jkl} \zeta_l + \zeta_j \varepsilon_{kil} \zeta_l) \tag{8.15e}$$

Because of the relation (8.14) there are definite relations between the scalar functions of (8.4) and those of (8.15). Some of these relations were obtained by CHANDRASEKHAR as follows:

$$H = - D A \tag{8.16a}$$

$$R = - D B \tag{8.16b}$$

$$U = - \left(\frac{\partial^2 E}{\partial r^2} + \frac{6}{r} \cdot \frac{\partial F}{\partial r}\right) \tag{8.16c}$$

$$V = - \left(\frac{\partial^2 F}{\partial r^2} + \frac{6}{r} \cdot \frac{\partial E}{\partial r}\right) \tag{8.16d}$$

$$P = - D G \tag{8.16e}$$

where D is the operator defined in equation (8.10).

The equations governing these scalar functions A, B, E, F, and G are

$$\frac{\partial A}{\partial t} = 2 \nu_H D A + 2 G + \text{constant} \tag{8.17}$$

$$\frac{\partial B}{\partial t} = \left(r \frac{\partial}{\partial r} + 5 \right) \cdot (E - F) + W + (\nu + \nu_H) D B + \text{constant} \tag{8.18}$$

5. The dissipation of energy. It is possible to expand these scalar functions in power series of r valid near $r = 0$. We have

$$\begin{aligned} Q &= Q_0 + Q_2 r^2 + \dots \\ H &= H_0 + H_2 r^2 + \dots \\ S &= S_0 + S_2 r^2 + \dots \\ T &= T_0 + T_2 r^2 + \dots \\ P &= P_0 + P_2 r^2 + \dots \end{aligned} \tag{8.19}$$

Substituting equation (8.19) into equations (8.10) and (8.11) and collecting the coefficients of the terms independent of r, we have

$$\frac{\mathrm{d} Q_0}{\mathrm{d} t} = - 10 S_0 + 20 \nu Q_2 \tag{8.20}$$

$$\frac{\mathrm{d} H_0}{\mathrm{d} t} = + 2 P_0 + 20 \nu_H H_2 \tag{8.21}$$

From equation (8.4) we have

$$\overline{u^2} = - 6 Q_0 ; \; \overline{h^2} = - 6 H_0 \tag{8.22}$$

and

$$P_0 = 5 S_0 = - \tfrac{5}{4} \overline{h_1^2 \frac{\partial u_1}{\partial x_1}} \tag{8.23}$$

Using the relations of equations (8.22) and (8.23), equations (8.20) and (8.21) become respectively

$$\tfrac{1}{2} \cdot \frac{\mathrm{d} \overline{u^2}}{\mathrm{d} t} = - 7.5 \overline{h_1^2 \frac{\partial u_1}{\partial x_1}} - 60 \nu Q_2 \tag{8.24}$$

$$\tfrac{1}{2} \cdot \frac{\mathrm{d} \overline{h^2}}{\mathrm{d} t} = + 7.5 \overline{h_1^2 \frac{\partial u_1}{\partial x_1}} - 60 \nu_H H_2 \tag{8.25}$$

Adding equations (8.24) and (8.25) gives

$$\tfrac{1}{2} \cdot \frac{\mathrm{d}}{\mathrm{d} t} (\overline{u^2} + \overline{h^2}) = - 60 (\nu Q_2 + \nu_H H_2) \tag{8.26}$$

Equation (8.26) shows that the rate of loss of energy of the system is due to viscous dissipation and Joule heat.

6. Invariants of Loitsiansky's type. Multiplying equation (8.10) by r^4 and integration from 0 to r, we have

$$\frac{\partial}{\partial t}\int_0^r Q\, r^4\, \mathrm{d}\, r = 2\, r^5\, (T - S) + 2\, \nu\, r^4 \frac{\partial Q}{\partial r} \tag{8.27}$$

If

$$r^5\, (T - S) \to 0 \text{ and } r^4\, Q' \to 0 \text{ as } r \to \infty \tag{8.28}$$

then

$$\int_0^\infty Q\, r^4 \cdot \mathrm{d}\, r = \text{constant} \tag{8.29}$$

Equation (8.29) is known as the Loitsiansky's invariant. It is interesting to notice that the Loitsiansky's invariant of ordinary turbulence holds true also in magnetohydrodynamics.

Similarly from equation (8.11) we have:

$$r^4\, G' \to 0,\ r^4\, H' \to 0,\ r^4\, A' \to 0 \text{ as } r \to \infty \tag{8.30}$$

then

$$\int_0^\infty r^4\, H\, \mathrm{d}\, r = 0 \tag{8.31}$$

From equation (8.12), we have:

if

$$r^5\, (U - V) \to 0,\ r^4\, W' \to 0,\ r^4\, R' \to 0,\ r^4\, B' \to 0 \text{ as } r \to \infty \tag{8.32}$$

then

$$\int_0^\infty R\, r^4\, \mathrm{d}\, r = 0 \tag{8.33}$$

7. A spontaneous magnetic field in a conducting fluid. Batchelor (*1*) discussed in detail the behavior of a spontaneous magnetic field in a turbulent flow without external magnetic and electric fields. In this case, the electromagnetic energy in the flow field must arise from the hydrodynamic motion. It is interesting to know how such a spontaneous magnetic field varies in turbulent motion. Some conclusions may be obtained if we assume that the disturbing electromagnetic fields are stationary random functions. From equation (8.6) we have the time rate of variation of the average magnetic energy in unit volume as follows:

$$\frac{1}{2} \cdot \frac{\partial \overline{|H|^2}}{\partial t} = \overline{H_i\, H_j \frac{\partial u_i}{\partial x_j}} - \overline{H_j\, u_j \frac{\partial H_i}{\partial x_j}} + \nu_H\, \overline{H_i \frac{\partial^2 H_i}{\partial x_j\, \partial x_j}} \tag{8.34}$$

If we assume that $\vec{H}$ is a stationary random function, we have

$$\frac{1}{2} \cdot \frac{\mathrm{d}\overline{H_i^2}}{\mathrm{d}t} = \overline{H_i^2 \left(\frac{\partial \vec{u}}{\partial x_H}\right)_H} - \nu_H\, \overline{\frac{\partial H_i}{\partial x_j} \cdot \frac{\partial H_i}{\partial x_i}} \tag{8.35}$$

where subscript H refers to the components in the direction of the line of magnetic force. Because of the analogy of $\vec{H}$ and $\vec{\omega}$ in ordinary hydrodynamics (see chapter IV, § 5), we have the corresponding equation of vorticity in ordinary homogeneous turbulence as follows:

$$\frac{1}{2} \cdot \frac{\mathrm{d}\, \overline{\omega_i^2}}{\mathrm{d}\, t} = \overline{\omega_i^2 \left(\frac{\partial \vec{u}}{\partial x_\omega}\right)_\omega} - \nu\, \overline{\frac{\partial \omega_i}{\partial x_j} \cdot \frac{\partial \omega_i}{\partial x_j}} \tag{8.36}$$

The first terms on the right hand side of equations (8.35) and (8.36) represent the terms that produce turbulence by extension and contraction of the lines of magnetic force and vorticity respectively. The second terms of these equations represent the dissipation terms. If we assume that the statistical properties of $\vec{H}$ and $\vec{\omega}$ are the same, we may apply some of the known results of $\vec{\omega}$ in an ordinary homogeneous turbulence to $\vec{H}$ in magnetohydrodynamics. For ordinary homogeneous turbulent flow, the two terms in the right hand side of equation (8.36) are approximately equal. Hence if $\nu = \nu_H$, the two terms on the right hand sid(of equation (8.35) will be also approximately equal. Thus we have that the criterion for an increase of the energy of the disturbance magnetic field is $\nu_H < \nu$, i. e.,

$$\mu_e \sigma \nu > 1 \tag{8.37}$$

Equation (8.35) may be written as follows:

$$\frac{1}{2} \cdot \frac{d \log \overline{H_i^2}}{dt} = \frac{\overline{H_i^2 \left(\frac{\partial \vec{u}}{\partial x_H}\right)_H}}{\overline{H_i^2}} - \nu_H \frac{\overline{\left(\frac{\partial H_i}{\partial x_j}\right)^2}}{\overline{H_i^2}} \tag{8.38}$$

If the statistical characteristics of $\vec{H}$ do not change appreciably with time except its magnitude, the right hand side of equation (8.38) is approximately constant. Thus $\overline{H_i^2}$ increases exponentially with time if $\nu_H < \nu$.

8. Stationary turbulence. Let us consider the stationary condition in which an external agency supplies kinetic energy to the system at a constant rate ε. Equations (8.24) and (8.25) become respectively

$$\varepsilon = 7.5\, \overline{h_1^2 \frac{\partial u_1}{\partial x_1}} + 60\, \nu\, Q_2 \tag{8.39}$$

$$0 = -7.5\, \overline{h_1^2 \frac{\partial u_1}{\partial x_1}} + 60\, \nu_H\, H_2 \tag{8.40}$$

where Q_2 and H_2 are now constants independent of time.

Equation (8.26) become

$$\varepsilon = 60\,(\nu\, Q_2 + \nu_H\, H_2) \tag{8.41}$$

Under stationary condition, equation (8.10) must be replaced by

$$\tfrac{1}{6}\,\varepsilon = \left(r \frac{d}{dr} + 5\right) \cdot (T - S) + \frac{\nu}{r^4} \cdot \frac{d}{dr}\left(r^4 \frac{dQ}{dr}\right) \tag{8.42}$$

Integration of equation (8.42) gives

$$\tfrac{1}{30}\,\varepsilon\, r = r\,(T - S) + \nu \frac{dQ}{dr} \tag{8.43}$$

For large Reynolds numbers, we may conclude that as $r \to \infty$

$$(T - S) \to \frac{\varepsilon}{30} \tag{8.44}$$

or

$$\overline{(u_1{}^2 + 2\,h_2{}^2)\,u_1{}'} \to -\tfrac{2}{15}\,\varepsilon\,r \tag{8.45}$$

Similarly under stationary condition, equation (8.11) must be replaced by

$$0 = P + \frac{\nu_H}{r^4}\cdot\frac{\mathrm{d}}{\mathrm{d}\,r}\left(r^4\frac{\mathrm{d}\,H}{\mathrm{d}\,r}\right) \tag{8.46}$$

Integration of equation (8.46) gives

$$\nu_H\,H - G = \text{constant}$$

Under stationary condition, we may set

$$\frac{\partial\,R}{\partial\,t} = \tfrac{1}{3}\,\xi = \text{constant} \tag{8.47}$$

Equation (8.12) becomes

$$\tfrac{1}{3}\,\xi = \left(r\frac{\mathrm{d}}{\mathrm{d}\,r} + 5\right)\cdot(U - V) + \frac{1}{r}\cdot\frac{\mathrm{d}}{\mathrm{d}\,r}\left\{r^4\frac{\mathrm{d}}{\mathrm{d}\,r}[(\nu + \nu_H)\,R - W]\right\} \tag{8.48}$$

Integration of equation (8.48) gives

$$\tfrac{1}{15}\,\xi\,r = r\,(U - V) + (\nu + \nu_H)\frac{\mathrm{d}\,R}{\mathrm{d}\,r} - \frac{\mathrm{d}\,W}{\mathrm{d}\,r} \tag{8.49}$$

As $r \to \infty$,

$$(U - V) \to \tfrac{1}{15}\,\xi \tag{8.50}$$

If the Loitsiansky invariant for R should be zero, we have

$$U - V \to 0 \ (\text{as } r \to \infty) \tag{8.51}$$

Equation (8.51) suggest that under stationary conditions the energy in the magnetic field is contained in wave numbers which are small compared to those in which the kinetic energy is contained.

9. The partition of energy in hydromagnetic turbulence. As we discuss in § 7, in a medium of high electrical conductivity, turbulence will be accompanied by a spontaneous generation and a subsequent amplification of a magnetic field. In a state of equilibrium, the energy in the magnetic field will approach that in the velocity field. We may write

$$\frac{\mu_e}{2}\,\overline{H^2} = k\,\tfrac{1}{2}\,\rho\,\overline{u^2} \tag{8.52}$$

where k is a constant of the order of unity. CHANDRASEKHAR was able to show that in the inertial sub-range in which Kolmogoroff's hypotheses of similarity holds the constant k is 1.62647. The readers who are interested in the detailed derivation of this result should refer to reference *4*.

References

1. BATCHELOR, G. K.: On the Spontaneous Magnetic Field in Conducting Liquid in Turbulent Motion. Proc. Roy. Soc. Ldonon A-201$_I$ ·950, pp. 405–416.

2. CHANDRASEKHAR, S.: The Invariant Theory of Isotropic Turbulence in Magnetohydrodynamics. Proc. Roy. Soc. London A-204, 1950, pp. 435–449.

3. CHANDRASEKHAR, S.: The Invariant Theory of Isotropic Turbulence in Magnetohydrodynamics. Proc. Roy. Soc. A-207, 1951, pp. 301–306.
4. CHANDRASEKHAR, S.: The Partition of Energy in Hydromagnetic Turbulence. Annals of Physics, vol. 2, 1957, pp. 615–626.
5. CHANDRASEKHAR, S.: Hydromagnetic Turbulence I. A Deductive Theory and II. An Elementary Theory. Proc. Roy. Soc. London A-233, 29 Dec. 1955, pp. 322–350.
6. KRAICHNAN, R. H.: Statistical Mechanics of Stationary Homogeneous Hydromagnetic Turbulence. Res. Rep. No. MH-7, Inst. of Math. Sci., New York Univ., 1957.
7. PAI, S. I.: Viscous Flow II–Turbulent Flow. D. Van Nostrand Co., Inc., Princeton, N. J., 1957.

Chapter IX

One Dimensional Flow in Magnetogasdynamics

1. Introduction. In chapters VI to VIII, we considered the plasma as an incompressible fluid of constant density. Actually the density of the plasma changes considerably with the speed as well as with the temperature of the flow. For high-speed flow or large temperature gradient flow or both, the effect of compressibility, i. e., the change of density, must be considered in the study of the flow of plasma. In the next four chapters, we are going to study the flow of plasma including the effect of compressibility. There are many interesting phenomena due to this compressibility effect such as shock wave phenomena. In order to bring out the essential features of the flow of a compressible fluid, we shall discuss the one-dimensional case first. We shall be concerned only with the velocity along a stream tube and the accompanied variations in pressure, density, and temperature of the plasma as well as the magnetic field.

In the following analysis, the following assumptions are made:

(i) The flow of the plasma is parallel to the x-axis and has the component of velocity u only.

(ii) The magnetic field $\vec{H}$ is planar and perpendicular to the velocity field u. If the magnetic field $\vec{H}$ is parallel to u, the magnetic field will not affect the gasdynamic equations. Then there will be no interaction phenomenon.

(iii) All quantities are functions of space coordinate, x and time t only.

The fundamental equations of our problem specified above will be discussed in § 2. In §§ 3 to 7, we shall further simplify our problem by considering the steady case only, i. e., all variables are functions of space coordinate x only. This steady flow problem describes the shock wave structure in a plasma. In §§ 8 and 9, we shall discuss the one-dimensional unsteady flow.

In § 10, we shall discuss the steady flow through a nozzle under the influence of a magnetic field which would be of some engineering interest.

In the above analysis, we consider strictly the one-dimensional flow problem where the flow variables are independent of the y and z coordinates. For engineering applications, we often use the "approximate" one-dimensional flow analysis for the flow inside a nozzle or a duct in which we consider only the average values over the cross-section of the nozzle. We consider that the average values of the flow variables are functions of the coordinates along the axis of the nozzle only. Hence the small variation of the same variables with respect to y and z are permissible. In this way, we may discuss some more general flows in a nozzle than the one discussed in the strict one-dimensional flow cases. We shall discuss this type of flow in nozzle in § 11.

The one-dimensional flow analysis may be used to predict the performance of the so-called MHD (Magnetohydrodynamic) power generation which will be briefly described in § 12.

2. Fundamental equations of one-dimensional unsteady flow of magnetogasdynamics (*6*). Under the assumptions listed in the last section, the fundamental equations of our problem (4.28) to (4.32) are reduced into the following forms:

$$p = \rho R_p T \tag{9.1a}$$

$$\frac{\partial \rho}{\partial t} + \frac{\partial \rho u}{\partial x} = 0 \tag{9.1b}$$

$$\rho \frac{\partial u}{\partial t} + \rho u \frac{\partial u}{\partial x} + \frac{\partial p}{\partial x} + \mu_e H \frac{\partial H}{\partial x} - \frac{4}{3} \cdot \frac{\partial}{\partial x}\left(\mu \frac{\partial u}{\partial x}\right) = 0 \tag{9.1c}$$

$$\frac{\partial H}{\partial t} + \frac{\partial u H}{\partial x} = \frac{\partial}{\partial x}\left(\nu_H \frac{\partial H}{\partial x}\right) \tag{9.1d}$$

$$\rho \frac{\partial h_0}{\partial t} + \rho u \frac{\partial h_0}{\partial x} - \frac{\partial p}{\partial t} - \frac{4}{3} \cdot \frac{\partial}{\partial x}\left(\mu u \frac{\partial u}{\partial x}\right) - \frac{\partial}{\partial x}\left(\varkappa \frac{\partial T}{\partial x}\right) + \mu_e \frac{\partial H}{\partial x}\left(u H - \nu_H \frac{\partial H}{\partial x}\right) = 0 \tag{9.1e}$$

Our problem is to investigate the five unknowns u, H, p, T, and ρ from the five equations of (9.1). The general discussion of the system of equations (9.1) is still very complicated. We are going to discuss some simple cases of this system first.

3. Steady one-dimensional flow (*5*). First we consider the steady flow case. In this case, all the partial derivatives with respect to time vanish, and all the variables are functions of x only. As a result, we may integrate equations (9.1b) to (9.1e) with respect to x and obtain the following relations:

$$\rho u = \text{constant} = m \tag{9.2}$$

$$m u + p + \frac{\mu_e H^2}{2} - \frac{4}{3} \mu \frac{\mathrm{d} u}{\mathrm{d} x} = \text{constant} = m C_1 \tag{9.3}$$

$$u H - \nu_H \frac{\mathrm{d} H}{\mathrm{d} x} = \text{constant} = E \tag{9.4}$$

$$m h_0 - \frac{4}{3} \mu u \frac{\mathrm{d} u}{\mathrm{d} x} + E \mu_e H - \varkappa \frac{\mathrm{d} T}{\mathrm{d} x} = \text{constant} = m C_2 \tag{9.5}$$

Equation (9.5) may be rewritten with the help of equation (9.3) in the form

$$\frac{\varkappa}{m R_p} \cdot \frac{\mathrm{d} R_p T}{\mathrm{d} x} = c_V T - \frac{u^2}{2} + C_1 u - \frac{\mu_e H^2}{2 \rho} + \frac{E \mu_e H}{m} - C_2 \tag{9.6}$$

It is convenient to introduce the non-dimensional variables

$$\begin{gathered} \xi = \frac{u}{u_1}, \quad \tau = \frac{R_p T}{u_1^2}, \quad h = \frac{H}{\sqrt{2 \rho_1 \frac{u_1^2}{\mu_e}}} \\ h_1 = \frac{H_1}{\sqrt{2 \rho_1 \frac{u_1^2}{\mu_e}}}, \quad P = \frac{p_1}{\rho_1 u_1^2} = \frac{2}{\gamma M_1^2} \end{gathered} \tag{9.7}$$

where M is the Mach number and the subscript 1 refers to the values of a uniform state in which

$$\frac{\mathrm{d}u}{\mathrm{d}x} = \frac{\mathrm{d}H}{\mathrm{d}x} = \frac{\mathrm{d}T}{\mathrm{d}x} = 0 \tag{9.8}$$

In terms of the non-dimensional parameters of equation (9.7), the constants C_1, C_2, and E are

$$\frac{C_1}{u_1} = 1 + P + h_1^2$$

$$\frac{C_2}{u_1^2} = \tfrac{1}{2} + \frac{\gamma P}{\gamma - 1} + 2 h_1^2 \tag{9.9}$$

$$\frac{E}{u_1^2} = h_1 \sqrt{\frac{2 \rho_1}{\mu_e}}$$

In terms of these non-dimensional variables, equations (9.3), (9.4) and (9.6), become respectively

$$\tfrac{4}{3} \cdot \frac{\mu}{m} \xi \frac{\mathrm{d}\xi}{\mathrm{d}x} = \tau + \xi^2 - (1 + P + h_1^2 - h^2)\, \xi \tag{9.10}$$

$$\frac{\nu_H}{u_1} \cdot \frac{\mathrm{d}h}{\mathrm{d}x} = \xi h - h_1 \tag{9.11}$$

$$\frac{\varkappa}{m R_p} \cdot \frac{\mathrm{d}\tau}{\mathrm{d}x} = \frac{\tau}{\gamma - 1} - \frac{\xi^2}{2} + (1 + P + h_1^2 - h^2)\, \xi - \tfrac{1}{2} - \frac{\gamma P}{\gamma - 1} - 2\,(h_1^2 - h h_1) \tag{9.12}$$

Our problem now is to solve for ξ, τ, and h from equations (9.10) to (9.11). After these variables are known, the density may be obtained from equation (9.2) and the pressure, from (9.1a). It should be noticed that μ, ν_H, and $\varkappa$ are usually functions of τ.

4. Generalized Rankine-Hugoniot relations (*5*). Our first question is this: starting from a uniform state 1, is it possible that the flow may change to another uniform state 2 which is different from state 1? In other words, we desire to know whether more than one uniform state satisfies the constants C_1, C_2, and E of our flow field. To find such uniform state, we substitute the conditions (9.8) into equations (9.10) to (9.12) and obtain

$$\tau + \xi^2 - (1 + P + h_1^2 - h^2)\, \xi = 0 \tag{9.13}$$

$$\xi h - h_1 = 0 \tag{9.14}$$

$$\tau + (\gamma - 1) \cdot \left[-\frac{\xi^2}{2} + (1 + P + h_1^2 - h^2)\, \xi - \tfrac{1}{2} - \frac{\gamma P}{\gamma - 1} - 2\,(h_1^2 - h h_1) \right] = 0 \tag{9.15}$$

Equations (9.13) to (9.15) are the generalized Rankine-Hugoniot relations across a normal shock in a transverse magnetic field. The general three-dimensional Rankine-Hugoniot relations of a shock (*1*) in magneto-gasdynamics will be discussed in chapter XI.

Eliminating τ and h from equations (9.13) to (9.15), we have

$$(\xi - 1) \cdot \left[\xi^2 - \left(\frac{\gamma - 1}{\gamma + 1} + \frac{2\gamma P}{\gamma + 1} + \frac{2\gamma h_1^2}{\gamma + 1} \right) \xi - \frac{2(2-\gamma) h_1^2}{\gamma + 1} \right] = \tag{9.16}$$
$$= (\xi - 1) \cdot (\xi - \xi_2) \cdot (\xi + \xi_3) = 0$$

There are three roots of equation (9.16). The negative root $-\xi_3$ has no physical significance because it represents the magnitude of the velocity of our flow field which is a positive quantity. The root unity is a trivial solution which represents the original flow. The only interesting solution is the root ξ_2 which gives the velocity after the region of transition which in certain cases is the shock front. The expression for ξ_2 is

$$\xi = \tfrac{1}{2} \left(\frac{\gamma - 1}{\gamma + 2} + \frac{2\gamma P}{\gamma + 1} + \frac{2\gamma h_1^2}{\gamma + 1} \right) + \tfrac{1}{2} \sqrt{\left(\frac{\gamma - 1}{\gamma + 1} + \frac{2\gamma P}{\gamma + 1} + \frac{2\gamma h_1^2}{\gamma + 1} \right)^2 + \frac{8(2-\gamma) h_1^2}{\gamma + 1}} \tag{9.17}$$

When there is no external magnetic field, $h_1 = 0$, and equation (9.17) becomes the well known relation of normal shock in ordinary gasdynamics.

From equation (9.17) we see that if P is kept unchanged, ξ_2 increases with h_1. As h_1 reaches a critical value h_c, $\xi_2 = 1$, where

$$h_c^2 = \frac{1 - \gamma P}{2} = \tfrac{1}{2} \left(1 - \frac{1}{M_1^2} \right) \tag{9.18}$$

If h_1 is larger than h_c, ξ_2 is greater than unity. If we assume that the flow in front of a shock has a Mach number $M_1 > 1$, there will be no compression shock possible in this flow field when h_1 is equal to or greater than h_c. If $h_1 = h_c$, no transition region will occur in this flow field because the only possible uniform state is $\xi_2 = 1$. If $h_1 > h_c$, the solution $\xi = \xi_2$ is physically impossible because of the second law of thermodynamics, i. e., the entropy will decrease in adiabatic flow. As we shall show in § 8, the velocity of propagation of small disturbance in the present case is not the ordinary sound speed but the effective sound speed a_e which is defined as

$$a_e^2 = \gamma R_p T + \frac{\mu_e H^2}{\rho} \tag{9.19}$$

The proper Mach number in the present case should be $M_e = u/a_e$. It is easy to show that $h_1 = h_c$, $M_{e1} = 1$ and $h_1 > h_c$, $M_{e1} < 1$. Hence the argument of ordinary gasdynamics that the shock wave will not occur in subsonic flow may be applied here when $M_{e1} < 1$.

If we assume that the gas is inviscid, non-heat-conducting and of infinite electrical conductivity, i. e., $\mu = \varkappa = 0$ and $\sigma = \infty$, equations (9.13) to (9.15) hold for the whole flow field instead of uniform states only in the general case $\mu \neq 0$, $\varkappa \neq 0$ and $\sigma \neq \infty$. For the ideal plasma, a surface of discontinuity is allowable. We may consider that the two uniform states $\xi = 1$, and $\xi = \xi_2$ are joined by a surface of discontinuity, i. e., a shock front. For actual gases, $\mu \neq 0$, $\varkappa \neq 0$ and $\sigma \neq \infty$. The surface of discontinuity should be replaced by a transition region in which ξ, τ, and h gradually change from the state 1 to the final state 2. In ordinary gasdynamics, we know that this transition region is very thin and it may be considered as a surface of discontinuity for many practical problems. The question how large is the thickness of this transition region in magnetogasdynamics should be answered from the study of equations (9.10) to (9.12). We are going to discuss various cases of this transition regions in the following three sections.

5. Shock wave structure in an inviscid and non-heat conducting fluid of finite electrical conductivity (*5*). In the present case, we have $\mu = \varkappa = 0$ and $\sigma \neq \infty$. We must solve equations (9.11), (9.13), and (9.15) for ξ, τ, and h. If we eliminate τ from equations (9.13) and (9.15), we have

$$\gamma (h \xi)^2 - 2 (\gamma - 1) h_1 (h \xi) - (2 - \gamma) h_1^2 + K = 0 \tag{9.20}$$

where

$$K = \frac{\gamma + 1}{2} (\xi - 1) \cdot (\xi - \xi_2) \cdot (\xi + \xi_3) \tag{9.21}$$

From equation (9.20), we have

$$h = \frac{(\gamma - 1) h_1 \pm \sqrt{h_1^2 - \gamma K}}{\gamma \xi} = h(\xi) \tag{9.22}$$

In the present case, there is a simple relation between h and ξ.

Substituting equation (9.22) into equation (9.11) yields

$$\frac{\mathrm{d} h}{\mathrm{d} x_1} = \frac{1}{\gamma} (- h_1 + \sqrt{h_1^2 - \gamma K}) \tag{9.23}$$

where $x_1 = x\, u_1/\nu_H$. The plus sign is used in equation (9.23) because $\mathrm{d}\,h/\mathrm{d}\,x = 0$ at the end points $\xi = 1$ and $\xi = \xi_2$.

We are interested in the region $1 > \xi > \xi_2$. In this region K is always negative and $\mathrm{d}\,h/\mathrm{d}\,x_1$ is always positive and finite. We would like to ask whether a continuous solution is possible in the present case? This question may be answered if we consider $\mathrm{d}\,\xi/\mathrm{d}\,x_1$ instead of $\mathrm{d}\,h/\mathrm{d}\,x_1$. We have

$$\frac{\mathrm{d} \xi}{\mathrm{d} x_1} = \frac{\mathrm{d} \xi}{\mathrm{d} h} \cdot \frac{\mathrm{d} h}{\mathrm{d} x_1} = F(\xi) \tag{9.24}$$

where $\mathrm{d}\,h/\mathrm{d}\,x_1$ is given by equation (9.23) and is a function of ξ while

$$\frac{\mathrm{d} \xi}{\mathrm{d} h} = \frac{2 \gamma \xi^2 \sqrt{h_1^2 - \gamma K}}{- \gamma \xi \dfrac{\mathrm{d} K}{\mathrm{d} \xi} - 2 \sqrt{h_1^2 - \gamma K} \left[(\gamma - 1) h_1 + \sqrt{h_1^2 - \gamma K} \right]} \tag{9.25}$$

In the region of interest $(1 > \xi > \xi_2)$, $\sqrt{h_1^2 - \gamma K}$ is always positive, whereas $\mathrm{d}\,K/\mathrm{d}\,\xi$ is first positive near $\xi = 1$ and becomes negative near $\xi = \xi_2$. It is possible that for a certain value of ξ between 1 and ξ_2, the denominator of equation (9.25) becomes zero, we have $\mathrm{d}\,\xi/\mathrm{d}\,h = \infty$ and then $\mathrm{d}\,\xi/\mathrm{d}\,x_1 = \infty$. This means that we should have a discontinuous solution. The critical value of ξ_c at which $\mathrm{d}\,\xi/\mathrm{d}\,x_1 = \infty$ is given by the equation

$$\gamma \xi \frac{\mathrm{d} K}{\mathrm{d} \xi} + 2 \sqrt{h_1^2 - \gamma K} \left[(\gamma - 1) h_1 + \sqrt{h_1^2 - \gamma K} \right] = 0 \tag{9.26}$$

When h_1 is small, we may always find a value of $\xi = \xi_c$ where $1 > \xi_c > \xi_2$ that satisfies equation (9.26). We shall then have discontinuous solutions. For instance, as $h_1 \to 0$, we have

$$\xi_c = \sqrt{\frac{\gamma - 1 + 2\gamma P}{\gamma + 1}} = \sqrt{\xi_{2_{h_1 \to 0}}} \tag{9.27}$$

It is evident that $1 > \xi_c > \xi_2$. As h_1 increases from zero to h_c, the maximum value of $|-\mathrm{d}\,K/\mathrm{d}\,\xi|$ tends to zero while the second term of equation (9.26) tends to be $2\gamma h_1^2$. For a given value of P, there is a value of $h_1 = h_b$ above which there is no value of ξ between 1 and ξ_2 which satisfies equation (9.26). Hence for $h_1 > h_b$, we shall have a continuous solution for the present problem.

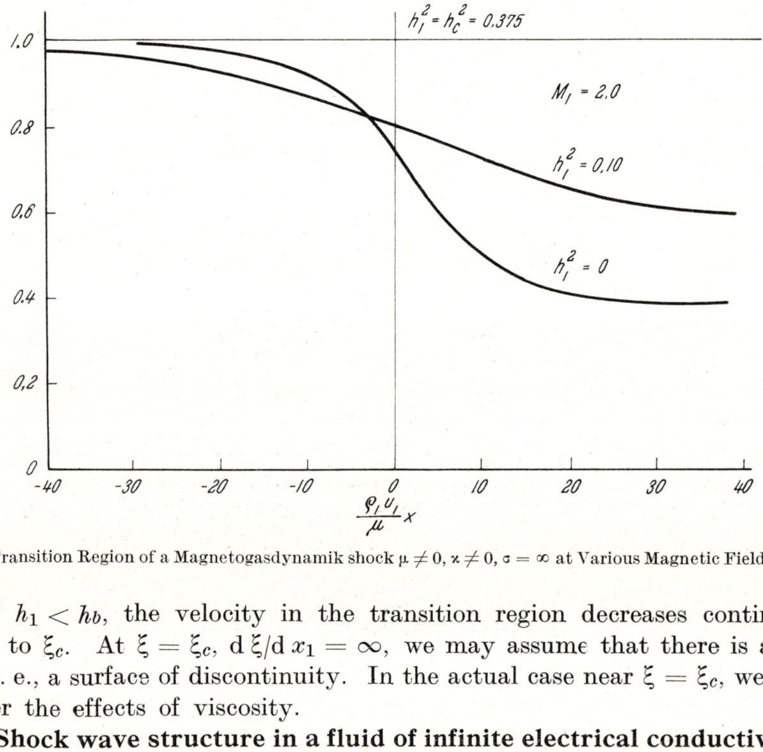

Fig. 9.1. Transition Region of a Magnetogasdynamik shock $\mu \neq 0$, $\varkappa \neq 0$, $\sigma = \infty$ at Various Magnetic Field Strengths

For $h_1 < h_b$, the velocity in the transition region decreases continuously from 1 to ξ_c. At $\xi = \xi_c$, $\mathrm{d}\,\xi/\mathrm{d}\,x_1 = \infty$, we may assume that there is a shock front, i. e., a surface of discontinuity. In the actual case near $\xi = \xi_c$, we should consider the effects of viscosity.

6. Shock wave structure in a fluid of infinite electrical conductivity (*5*). In this case, we have $\mu \neq 0$, $\varkappa \neq 0$ and $\sigma = \infty$. We should solve equations (9.10), (9.12), and (9.14) for ξ, τ, and h. Eliminating h and x from these three equations we obtain a first order differential equation for τ in terms of ξ which is

$$\frac{\mathrm{d}\,\tau}{\mathrm{d}\,\frac{\xi^2}{2}} = \frac{4}{3} \cdot \frac{P_r}{\gamma} \cdot \frac{\tau - \tau_0}{\tau - \tau_\infty} \tag{9.28}$$

where $P_r = \mu\, C_p/\varkappa =$ Prandtl number and

$$\tau_0 = P - (\gamma - 1) \cdot \left[-\frac{\xi^2}{2} + (1 + P + h_1^2)\,\xi - \left(\frac{1}{2} + P + 2\,h_1^2 - \frac{h_1^2}{\xi}\right)\right] \tag{9.29}$$

$$\tau_\infty = -\xi^2 + (1 + P + h_1^2)\,\xi - \frac{h_1^2}{\xi} \tag{9.30}$$

Regardless of whether or not the Prandtl number P_r is constant, if $0 < P_r < \infty$, equation (9.28) has a singularity whenever the numerator and the denominator of the right hand side of equation (9.28) vanish simultaneously. The conditions $\tau = \tau_0$ and $\tau = \tau_\infty$ lead to equation (9.16). Hence the two uniform states defined by equation (9.16) are two singular points of equation (9.28). We are interested in the integral curves of equation (9.28) in the $\xi - \tau$ plane joining the two uniform states $\xi = 1$ and $\xi = \xi_2$. Such integral curves may be obtained by means of isoclines. Since $\tau = \tau_0$, $\mathrm{d}\,\tau/\mathrm{d}\,(\xi^2/2) = 0$ and $\tau = \tau_\infty$, $\mathrm{d}\,\tau/\mathrm{d}\,\xi^2/2) = \infty$, the integral curve must lie between the two curves $\tau = \tau_0$ and $\tau = \tau_\infty$ in the $\xi - \tau$ plane.

From equation (9.28), we may obtain

$$\tau = \tau\,(\xi) \tag{9.31}$$

In general, numerical or graphical integrations should be used to obtain relation (9.31). After $\tau = \tau\,(\xi)$ is obtained, the x-coordinate may be obtained by simple quadrature, i. e.,

$$\frac{\rho_1 u_1}{\mu_1} \cdot x = \frac{4}{3} \int \frac{\dfrac{\mu}{\mu_1}}{\tau - \tau_\infty}\,\mathrm{d}\left(\frac{\xi^2}{2}\right) \tag{9.32}$$

Fig. 9.1 shows some typical curves for the transition regions at $M_1 = 2.0$, $\gamma = 1.4$, $P_r = 3/4$, $(\mu/\mu_1) = (\tau/\tau_1)^{0.76}$ and $h_1^2 = 0$, 0.1, and 0.375. The interesting point is that as h_1 increases from zero to h_c, the thickness of the transition region increases considerably. Of course, for $h_1 = h_c$, the thickness of transition region is infinite. When $h_1 > h_c$, no compression shock is possible.

7. Shock wave structure in a plasma. In an actual plasma, we have $\mu \neq 0$, $\varkappa \neq 0$ and $\sigma \neq \infty$. In this case, we should solve equations (9.10) to (9.12) for ξ, τ, and h between the two uniform states $\xi = 1$ and $\xi = \xi_2$. It is easy to show that these two uniform states also represent the singularity of this system of equations. Marshall discussed this case in reference *4*. Qualitatively, the flow field is a combination of the results of the last two sections. When the value of h_1 is below the lower critical value h_b, we expect that the velocity first decreases gradually from $\xi = 1$ to a value near ξ_c, then decreases rapidly to a value near ξ_2 and finally decreases gradually to ξ_2 as the flow proceeds. For h_1 lying between h_b and h_c, we would expect that the rate of decrease of velocity ξ is small in the whole flow field. We thus have a wide transition region and shall not have a sharp transition region as in a shock front. Finally for $h_1 > h_c$, no compression transition region nor shock front would occur, if we consider the state $\xi = 1$ as the flow in front of the transition region. The only possible solution is a uniform stream $\xi = 1$. As the velocity decreases, the magnetic field h increases. In the slow transition region, the magnetic field increases gradually while in the sharp transition region or shock front, the magnetic field increases rapidly.

8. Unsteady flow of an ideal plasma (*6*). For ideal plasma, $\mu = \varkappa = 0$ and $\sigma = \infty$ or $\nu_H = 0$. Equation (9.1) becomes

$$p = \rho R_p T \tag{9.33a}$$

$$\frac{\partial \rho}{\partial t} + u \frac{\partial \rho}{\partial x} = -\rho \frac{\partial u}{\partial x} \tag{9.33b}$$

$$\rho \frac{\partial u}{\partial t} + \rho u \frac{\partial u}{\partial x} + \frac{\partial p}{\partial x} + \mu_e H \frac{\partial H}{\partial x} = 0 \tag{9.33c}$$

$$\frac{\partial H}{\partial t} + u \frac{\partial H}{\partial x} = -H \frac{\partial u}{\partial x} \tag{9.33d}$$

$$\rho \frac{\partial h_0}{\partial t} + \rho u \frac{\partial h_0}{\partial x} - \frac{\partial p}{\partial t} + \mu_e u H \frac{\partial H}{\partial x} = 0 \tag{9.33e}$$

Since equations (9.33b) and (9.33d) are identical in form, we may write

$$H = A \rho = \frac{H_0}{\rho_0} \rho \tag{9.34}$$

where A is an arbitrary constant determined by the initial conditions and subscript 0 refers to the values of the stagnation point. The functional forms of H and ρ are the same.

From equations (9.33a), (9.33c), and (9.33e), we have

$$\left(\frac{\partial}{\partial t} + u \frac{\partial}{\partial x}\right) \ln \frac{p}{\rho^\gamma} = 0 \tag{9.35}$$

or

$$\left(\frac{\partial}{\partial t} + u \frac{\partial}{\partial x}\right) S = 0 \tag{9.35a}$$

This shows that along any line of flow, the entropy is constant. This relation is the same as that of ordinary gasdynamics of inviscid fluid. In both cases, there is no dissipation mechanism. One of the particular solutions of (9.35a) is $S =$ constant throughout the whole flow field, i. e., isentropic flow. However, in general, a shock wave might occur and the entropy, therefore, might change in the flow field.

Now we are going to find an exact solution of equation (9.35) such that the flow is isentropic ($S =$ constant) and such that the density is a function of velocity u only. We may write

$$\rho = f(u) \tag{9.36}$$

where the function f will be determined in the following manner:

Equation (9.33b) becomes

$$\frac{\partial u}{\partial t} + u \frac{\partial u}{\partial x} = -\frac{f}{f'} \cdot \frac{\partial u}{\partial x} \tag{9.37}$$

where $f' = \mathrm{d} f/\mathrm{d} u = \mathrm{d} \rho/\mathrm{d} u$. Equation (9.33c) becomes

$$\frac{\partial u}{\partial t} + u \frac{\partial u}{\partial x} = -\frac{a_e^2}{f} f' \frac{\partial u}{\partial x} \tag{9.38}$$

where a_e is the effective sound speed of the present problem and is defined by

$$a_e^2 = \gamma R_p T + \frac{\mu_e H^2}{\rho} \tag{9.39}$$

In the present case, a_e is a function of ρ only. In the general case a_e is a function of ρ and S.

If equation (9.36) is a solution of our problem, equations (9.37) and (9.38) must be the same. Hence we have

$$\frac{f'}{f} = \frac{1}{\rho} \cdot \frac{\partial \rho}{\partial u} = \pm \frac{1}{a_e} \tag{9.40}$$

Integration of equation (9.40) gives

$$u = \pm \int_{\rho_0}^{\rho} \frac{a_e(\rho)}{\rho} d\rho \tag{9.41}$$

If we substitute equation (9.40) into equation (9.37), we have

$$\begin{aligned} \frac{\partial u}{\partial t} + (u \pm a_e) \frac{\partial u}{\partial x} &= 0 \\ \frac{\partial \rho}{\partial t} + (u \pm a_e) \frac{\partial \rho}{\partial x} &= 0 \end{aligned} \tag{9.42}$$

It is easy to show that the general solutions of equation (9.42) are

$$u = F_1 [x - (u \pm a_e) t], \quad \rho = F_2 [x - (u \pm a_e) t] \tag{9.43}$$

where F_1 and F_2 are arbitrary functions. To show the significance of equation (9.43), let us consider the solution

$$u = F_1 [x - (u + a_e) t] \tag{9.44}$$

This solution means that the variation of u with respect to a point moving with velocity $u + a_e$ is zero, i. e., the disturbance is propagated with an instantaneous velocity $u + a_e$. If the velocity u is much smaller than a_e, we have $u = F_1 (x - a_e t)$. Thus a_e is the velocity of propagation of an infinitesimal disturbance in the present case. Hence we call a_e the effective speed of sound. For waves of finite amplitude, the velocity of propagation is different at different points in the flow field, and the shape of the wave will be distorted as the wave propagates. For waves of the type of equation (9.44), the velocity of propagation at the crest is larger than that at the trough. Hence the crest will overtake the trough. As a result, a shock wave will form and the assumption of isentropic flow breaks down. Qualitatively, the manner of formation of a shock wave in magnetogasdynamics is the same as that in ordinary gasdynamics. But since a_e is larger than a, the shock wave formation will be faster in magnetogasdynamics.

The solution $u = F_1 [x - (u - a_e) t]$ represents an expansion wave while $F_1 [x - (u + a_e) t]$ represents a compression wave. The expansion wave will smooth itself out as the wave propagates while the compression wave will form a shock as the wave propagates.

The system of the fundamental equations (9.33) is of hyperbolic type. We may use the well known method of characteristics to solve such problems (*2*). There are four sets of characteristic curves for equation (9.33), i. e., double roots of

$$\frac{\partial x}{\partial t} = u, \quad \text{flow lines}$$

and (9.45)

$$\frac{\partial x}{\partial t} = u \pm a_e, \quad \text{characteristic lines}$$

Along the flow lines $d\,x/d\,t = u$, we have

$$\frac{H}{\rho} = \text{constant} \tag{9.46}$$

and

$$S = \text{constant} \tag{9.47}$$

Along the characteristic lines $d\,x/d\,t = u \pm a_e$, we have respectively (*3*)

$$\mu_e H\left[\frac{\partial H}{\partial t} + (u \pm a_e)\frac{\partial H}{\partial x}\right] \pm \rho\, a_e\left[\frac{\partial u}{\partial t} + (u \pm u_e)\frac{\partial u}{\partial x}\right] + \left[\frac{\partial p}{\partial t} + (u \pm a_e)\frac{\partial p}{\partial x}\right] = 0 \tag{9.48}$$

The well-known numerical method of characteristics may be used to solve the one dimensional unsteady flow problems of magnetogasdynamics with the help of relations (9.45) to (9.48).

9. Characteristics of the fundamental equations of one-dimensional unsteady flow of magnetogasdynamics (*6*). The general equation (9.1) may be rewritten in the following form:

$$\rho\left(\frac{\partial u}{\partial t} + u\frac{\partial u}{\partial x}\right) + p_S\frac{\partial S}{\partial x} + p_\rho\frac{\partial \rho}{\partial x} + \mu_e H\frac{\partial H}{\partial x} - \tfrac{4}{3}\cdot\frac{\partial \tau}{\partial x} = 0 \tag{9.49a}$$

$$\rho\frac{\partial u}{\partial x} + \left(\frac{\partial \rho}{\partial t} + u\frac{\partial \rho}{\partial x}\right) = 0 \tag{9.49b}$$

$$H\frac{\partial u}{\partial x} + \left(\frac{\partial H}{\partial t} + u\frac{\partial H}{\partial x}\right) - \frac{\partial N}{\partial x} = 0 \tag{9.49c}$$

$$\rho\, T\left(\frac{\partial S}{\partial t} + u\frac{\partial S}{\partial x}\right) - \frac{\partial J}{\partial x} = \tfrac{4}{3}\cdot\frac{\tau^2}{\mu} + \frac{\mu_e}{\nu_H}N^2 \tag{9.49d}$$

$$\mu\frac{\partial u}{\partial x} = \tau \tag{9.49e}$$

$$\nu_H\frac{\partial H}{\partial x} = N \tag{9.49f}$$

$$\varkappa\frac{\partial T}{\partial x} = \varkappa\, T_S\frac{\partial S}{\partial x} + \varkappa\, T_\rho\frac{\partial \rho}{\partial x} = J \tag{9.49g}$$

The equation of state may be written as

$$p = p(\rho, S) \quad \text{or} \quad T = T(\rho, S) \tag{9.50}$$

Then

$$\begin{aligned} p_S &= \left(\frac{\partial p}{\partial S}\right)_\rho, \quad p_\rho = \left(\frac{\partial p}{\partial \rho}\right)_S \\ T_S &= \left(\frac{\partial T}{\partial S}\right)_\rho, \quad T_\rho = \left(\frac{\partial T}{\partial \rho}\right)_S \end{aligned} \tag{9.51}$$

Equation (9.49) may be considered as a system of seven first order partial differential equations for the unknowns u, ρ, H, S, τ, N, and J and the independent variables x and t. If $\psi(x, t) =$ constant is the equation of the characteristics of the system of equations, the characteristic condition is given by the following determinantal equation:

$$\begin{vmatrix} \rho\psi_t + \rho u \psi_x & p_\rho \psi_x & \mu_e H \psi_x & p_S \psi_x & -\frac{4}{3}\psi_x & 0 & 0 \\ \rho\psi_x & \psi_t + u\psi_x & 0 & 0 & 0 & 0 & 0 \\ H\psi_x & 0 & \psi_t + u\psi_x & 0 & 0 & -\psi_x & 0 \\ 0 & 0 & 0 & \rho T(\psi_t + u\psi_x) & 0 & 0 & -\psi_x \\ \mu\psi_x & 0 & 0 & 0 & 0 & 0 & 0 \\ 0 & 0 & \nu_H \psi_x & 0 & 0 & 0 & 0 \\ 0 & \varkappa T_\rho \psi_x & 0 & \varkappa T_S \psi_x & 0 & 0 & 0 \end{vmatrix} = 0 \tag{9.52}$$

where $\psi_t = \partial\psi/\partial t$ and $\psi_x = \partial\psi/\partial x$. From equation (9.52) we obtain the characteristic conditions of our problem as follows:

(1) If $\mu \neq 0$, $\varkappa \neq 0$ and $\nu_H \neq 0$, equation (9.52) becomes

$$\tfrac{4}{3}\mu\varkappa\nu_H T_S \psi_x^6 (\psi_t + u\psi_x) = 0 \tag{9.53}$$

If one or more of the values of μ, $\varkappa$, and ν_H are zero, the characteristic condition will be different from that of equation (9.53). The characteristic conditions for the following seven simple cases have been obtained:

(2) If $\mu = 0$, $\varkappa \neq 0$, $\nu_H \neq 0$, the characteristic condition is

$$\nu_H \varkappa \rho T_S \psi_x^4 [\psi_t + (u + a_T)\psi_x] \cdot [\psi_t + (u - a_T)\psi_x] = 0 \tag{9.54}$$

where $a_T = \sqrt{RT} =$ isothermal speed of sound.

(3) If $\mu \neq 0$, $\varkappa = 0$, $\nu_H \neq 0$, the characteristic condition is

$$\tfrac{4}{3}\mu\nu_H \rho T_S \psi_x^4 (\psi_t + u\psi_x)^2 = 0 \tag{9.55}$$

(4) If $\mu \neq 0$, $\varkappa \neq 0$, $\nu_H = 0$, the characteristic condition is

$$\tfrac{4}{3}\varkappa\mu T_S \psi_x^4 (\psi_t + u\psi_x)^2 = 0 \tag{9.56}$$

(5) $\mu = \varkappa = 0$, $\nu_H \neq 0$, the characteristic condition is

$$\rho^2 T_S \nu_H \psi_x^2 (\psi_t + u\psi_x) \cdot [\psi_t + (u + a)\psi_x] \cdot [\psi_t + (u - a)\psi_x] = 0 \tag{9.57}$$

where $a = \sqrt{p_\rho} =$ isentropic speed of sound.

(6) $\mu = \nu_H = 0$, $\varkappa \neq 0$, the characteristic condition is

$$\varkappa T_S \rho \psi_x^2 (\psi_t + u\psi_x) \cdot [\psi_t + (u + a_e)\psi_x] \cdot [\psi_t + (u - a_e)\psi_x] = 0 \tag{9.58}$$

where $a_e = \sqrt{(\mu_e H^2)/\rho + a^2} =$ effective speed of sound.

(7) $\varkappa = \nu_H = 0$, $\mu \neq 0$, the characteristic condition is

$$\tfrac{4}{3} \mu \rho T_S \psi_x^2 (\psi_t + u \psi_x)^3 = 0 \tag{9.59}$$

(8) $\mu = \varkappa = \nu_H = 0$, the characteristic condition is

$$\rho^2 T_S (\psi_t + u \psi_x)^2 \cdot [\psi_t + (u + a_e) \psi_x] \cdot [\psi_t + (u - a_e) \psi_x] = 0 \tag{9.60}$$

This is the case of an ideal plasma which we discussed in § 8.

From the above eight different cases, we see that there are three different kinds of characteristics in the one-dimensional unsteady flow of magnetogasdynamics. The first kind of characteristics is

$$\psi_t + u \psi_x = 0, \text{ or } \frac{dx}{dt} = -\frac{\psi_t}{\psi_x} = u \tag{9.61}$$

This is the line of flow which occurs in all the eight cases.

The second kind of characteristics is

$$\psi_x = 0 \quad \text{or} \quad t = \text{constant} \tag{9.62}$$

This is the parabolic characteristic, where the signal speed of disturbance is infinite. Except for case (8), this characteristic exists in all the other cases. If any one of the diffusion properties μ, $\varkappa$, or ν_H is different from zero, the parabolic characteristic exists.

The third kind of characteristics may be written in the following form:

$$\psi_t + (u \pm V_i) \psi_x = 0, \quad \text{or} \quad \frac{dx}{dt} = -\frac{\psi_t}{\psi_x} = u \pm V_i \tag{9.63}$$

where V_i may be either a_T, a, or a_e. These characteristics are lines of propagation of small disturbances with velocity of V_i relative to the fluid motion. These characteristics occur only in an inviscid fluid, i. e., $\mu = 0$. The value of V_i depends on the properties of the fluid. From the above results we have

(i) $\mu = 0$, $\varkappa \neq 0$, $\nu_H \neq 0$, $V_i = a_T$

(ii) $\mu = 0$, $\varkappa = 0$, $\nu_H \neq 0$, $V_i = a$

(iii) $\mu = 0$, $\varkappa = 0$, $\nu_H = 0$, or $\mu = \nu_H = 0$, $\varkappa \neq 0$, $V_i = a_e$

It may be shown that V_i is essentially the group velocity of propagation of waves as $\mu \to 0$.

10. Steady flow through a nozzle. Now we consider the steady flow of plasma through a nozzle which is of engineering interest. If the variation of the cross-sectional area of the nozzle is small, we may consider the flow as one-dimensional so that all the flow variables are functions of x, the coordinate along the axis of the nozzle, only. For large Reynolds number, we may neglect the viscosity and heat conductivity in our analysis. We shall, however, consider the electrical conductivity σ to be arbitrary. We shall assume for simplicity that there is a uniform external applied magnetic field in the y-direction which is perpendicular to the flow direction x such that $H_y = H_0$. The total magnetic field strength is then $H(x) = H_0 + h(x)$ where $h(x)$ is the induced magnetic field due to the flow. There is also an uniform external applied electric field in the z-direction which is perpendicular to both the x- and y-axis such that $E_{z0} = E_0$. The total electric field is $E(x) = E_0 + e(x)$ where $e(x)$ is the induced electric field due to the flow. The are definite relations between E and H and between E_0 and H_0 as given by equation (4.22). According to our discussion in chapter IV, § 6, our fundamental equations of the present problem are still those of magneto-gasdynamics which are:

(i) Equation of continuity: (4.2)

$$\rho\, u\, A = \text{constant} \tag{9.64}$$

where A is the cross sectional area of the nozzle which is in general a function of x, i. e., $A = A(x)$. In any given problem, $A(x)$ is assumed to be given. Logarithmic differentiation of equation (9.64) gives

$$\frac{\mathrm{d}A}{A} + \frac{\mathrm{d}u}{u} + \frac{\mathrm{d}\rho}{\rho} = 0 \tag{9.65}$$

(ii) Equation of motion: (4.25)

$$\rho\, u \frac{\mathrm{d}u}{\mathrm{d}x} + \frac{\mathrm{d}p}{\mathrm{d}x} = -\mu_e H \frac{\mathrm{d}H}{\mathrm{d}x} \tag{9.66}$$

where the viscous force is neglected in the present problem.

(iii) Equation of energy: (4.26)

$$\rho\, u\, C_p \frac{\mathrm{d}T}{\mathrm{d}x} + \rho\, u^2 \frac{\mathrm{d}u}{\mathrm{d}x} = \mu_e \frac{\mathrm{d}H}{\mathrm{d}x}\left(\nu_H \frac{\mathrm{d}H}{\mathrm{d}x} - u\, H\right) \tag{9.67}$$

where both the heat loss due to heat conduction and the loss due to viscous dissipation are neglected.

(iv) Equation of magnetic field: (4.24)

$$\frac{\mathrm{d}\, u H}{\mathrm{d}x} = \frac{\mathrm{d}}{\mathrm{d}x}\left(\nu_H \frac{\mathrm{d}H}{\mathrm{d}x}\right) \tag{9.68}$$

or

$$\nu_H \frac{\mathrm{d}H}{\mathrm{d}x} = u\, H + E_0 \tag{9.68a}$$

where E_0 is the constant of integration which depends on the external applied electric field.

(v) Equation of state: (4.1)

$$\frac{\mathrm{d}p}{p} = \frac{\mathrm{d}\rho}{\rho} + \frac{\mathrm{d}T}{T} \tag{9.69}$$

Now we have to solve the five unknowns u, H, p, ρ, and T from the five equations (9.65) to (9.69). We shall consider the following three cases:

(a) The case of infinite electrical conductivity $\sigma = \infty$, or $\nu_H = 0$. We shall consider first isentropic flow only because from § 8, we know that $S = \text{constant}$ is a solution of the present problem and the pressure is a function of density only. Hence the ordinary sound speed $a = \sqrt{\gamma p/\rho}$ is also a function of density only. In the present case, equation (9.68a) may be replaced by

$$\frac{d u}{u} + \frac{d H}{H} = 0 \tag{9.70}$$

It is then easy to show that from equation (9.65) to (9.69) with $\nu_H = 0$, we have

$$\frac{d u}{d x} = \frac{u a^2}{(u^2 - a_e^2)} \cdot \frac{1}{A} \cdot \frac{d A}{d x} \tag{9.71}$$

where a_e is the effective sound speed defined by equation (9.39). Equation (9.71) is of exactly the same form as the corresponding equation for ordinary gasdynamics except we have replaced a by a_e in the denominator. Hence we may draw the same conclusions: In particular, at the neck of the nozzle $d A/d x = 0$, the maximum velocity is the effective sound speed. The flow field may be divided into two regions, i. e., the venturi tube region in which the maximum speed is the effective sound speed and the flow is always isentropic and the de Laval nozzle region in which the flow effective Mach number $M_e = u/a_e$ may be larger or smaller than unity and a shock may occur in this region. The main difference of the present case from that of ordinary gasdynamics is that we have to use the effective Mach number in place of the ordinary Mach number. As we showed in the last section, the speed of propagation of small disturbances in the present case is a_e [cf. (9.60)].

(b) The case of finite electrical conductivity $\sigma \neq \infty$, or $\nu_H \neq 0$. We have to use equations (9.65) to (9.69) to solve our problem. It should be pointed out that in this case the speed of propagation of small disturbances is the ordinary sound speed a [cf. (9.57)]. Then the ordinary Mach number $M = u/a$ is the characteristic number for the present problem instead of M_e. From our fundamental equations (9.65) to (9.69) we have the following relations:

$$\frac{d u}{d x} = \frac{1}{M^2 - 1} \left[\frac{u}{A} \cdot \frac{d A}{d x} - \frac{\mu_e^2 \sigma}{p} (E_0 + u H) \cdot \left(u H + \frac{\gamma - 1}{\gamma} E_0\right)\right] \tag{9.72}$$

$$\nu_H \frac{d H}{d x} = u H + E_0 \tag{9.68a}$$

We may integrate the above two equations from the given initial conditions to obtain the distributions of u and H in the nozzle. It should be pointed out that in equation (9.72) the factor $(M^2 - 1)$ appears instead of the factor $(M_e^2 - 1)$ in equation (9.71) as we would expect. The difference of equation (9.72) from the corresponding equation in ordinary gasdynamics is the second term in the right-hand side of equation (9.72), which should be solved simultaneously with equation (9.68a). One general conclusion is that the acceleration of the flow $d u/d x$ changes sign at $M = 1$ for a given electromagnetic field. It is not necessary that the maximum speed at the neck be sonic because of the effect of the second term due to the electromagnetic field in the right-hand side of equation (9.72).

(c) The case of very small electrical conductivity, i. e., small Magnetic Reynolds number R_σ. In this case, we have shown in chapter IV, § 4 that the induced electromagnetic fields $h(x)$ and $e(x)$ are negligible because the term $u H$ is negligible in comparison with the other two terms in equation (9.68a). Hence we may assume $H(x) = H_0$ and $E(x) = E_0$ and discard equation (9.68a). The second term in the right-hand side of equation (9.72) may be considered than as a given function of x. This case has been discussed by RESLER and SEARS in reference *8*.

11. Approximate one-dimensional flow in a transverse electromagnetic field. In §§ 2 to 10 we consider the strict one-dimensional flow in which the z-component of the electric field $E_z = E$ should be a constant and the y-component of the magnetic field depends on the flow velocity as given by equation (9.68a). However in certain cases we may relax these conditions by considering the "quasi-one-dimensional" magnetogasdynamics in which we assume the following conditions to be fulfilled:

(i) The channel area changes slowly along its length. Thus we may discuss the mean values of the fluid properties at each section x of the channel.

(ii) We impose a transverse magnetic field $H_y = H(x)$ and a transverse electric field $E_z = E(x)$ along the boundary of the channel. We assume that the variations of $H(x)$ and $E(x)$ with x are not large so that the components H_x and E_x which are needed to satisfy the Maxwell equations are negligible. Furthermore, we assume that the magnetic Reynolds number is not too large so that the induced electromagnetic field strengths are negligible and that we may consider $H(x)$ and $E(x)$ hold for the whole section rather than at the boundary only. Under these conditions, the fundamental equations for one-dimensional flow through a channel of cross-section area $A(x)$ with the imposed electromagnetic fields $H(x)$ and $E(x)$ are

$$\rho u A = \text{constant} \tag{9.73a}$$

$$\rho u \frac{d u}{d x} + \frac{d p}{d x} = \sigma [E(x) + u \mu_e H(x)] \mu_e H(x) \tag{9.73b}$$

$$C_p \rho u \frac{d T}{d x} + \rho u^2 \frac{d u}{d x} = \sigma [E(x) + u \mu_e H(x)] E(x) + Q \tag{9.73c}$$

$$p = \rho R T \tag{9.73d}$$

where Q denotes heat added other than the joule heat such as those due to chemical reactions or viscous dissipations.

The main difference between equation (9.73) and those of § 10 is that $E(x)$ and $H(x)$ are now arbitrary functions of x. From strict one-dimensional analysis this equation (9.73) can be deduced only when the magnetic Reynolds number is very small and then E must be a constant. However if the conditions stated in this section are fulfilled, equation (9.73) may be considered as a good approximation for the actual case and the results will be more general than those given in § 10.

From equation (9.73) we have

$$\frac{d u}{d x} = \frac{\frac{u}{A} \cdot \frac{d A}{d x} - \frac{\gamma - 1}{\gamma} \cdot \frac{Q}{p} - \frac{\sigma \mu_e^2 H^2}{p} \left(u + \frac{E}{\mu_e H}\right) \cdot \left(u + \frac{\gamma - 1}{\gamma} \cdot \frac{E}{\mu_e H}\right)}{M^2 - 1} \tag{9.74}$$

where $M = u/a$ is the ordinary Mach number of the flow. As we have shown in section 10, for finite electric conductivity the ordinary Mach number is the critical parameter. Without the electromagnetic fields E and H, equation (9.74) is identical with those quasi-non-dimensional equation of gasdynamics with heat addition. In a way the electromagnetic fields have a similar effects as the heat addition. Equation (9.74) has been extensively discussed by SEARS and RESLER in reference *9*.

12. Magnetohydrodynamic (MHD) power generation. The basic principle of the MHD generator is the use of the motion of a fluid conductor in the presence of a magnetic field to generate the electric power. This idea had been known to FARADAY in 1831 who experimented with mercury flowing in a glass tube between the poles of a magnet and proposed use of tidal currents in the terrestrial magnetic field for power generation. Only recently experiments with

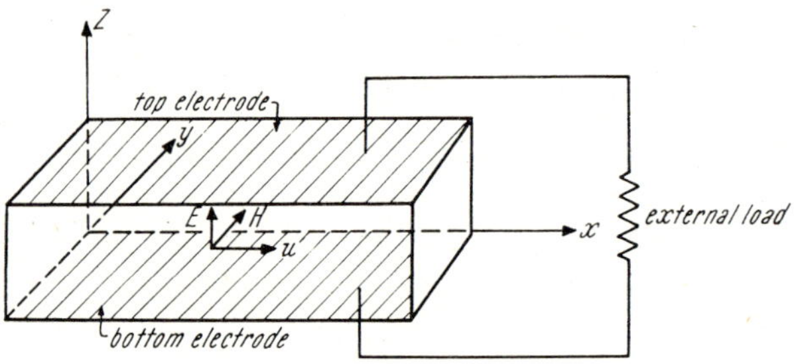

Fig. 9.2. A sketch of MHD generator

ionized gas as the conducting fluid have been extensively carried out (*10, 11, 12*). It has been found that for high power application such as the power over one megawatt the MHD generator seems to be more compact and lighter than other devices. At the present time, the MHD generator is still in the development stage because of the lack of sufficient knowledge in high temperature technology.

Fig. 9.2 shows a sketch of a protol type MHD generator which consists of a channel with top and bottom electrodes. The ionized gas flows in the channel with a mean velocity u over a section x of the channel. A uniform magnetic field H_y is applied in the y-direction. Electric current I is flowing from the top electrode through an external load to the bottom electrode. It is evident that our one-dimensional analysis discussed above may be used to predict the performance of the MHD generator. Experimental results check the theoretical prediction very well (*10*).

References

1. DE HOFFMANN, F., and E. TELLER: Magneto-Hydrodynamic Shock. Phys. Rev., vol. 80, No. 4, 1950, pp. 692–703.

2. FRIEDRICHS, K. O.: Nonlinear Wave Motion in Magneto-Hydrodynamics. Los Alamos Laboratory Report, 1954.

3. LOUGHHEAD, R. E.: Solution of Problems Involving in Hydromagnetic Flow of Compressible Ionized Fluids. Phys. Rev., vol. 99, No. 6, 1955, pp. 1678–1681.

4. MARSHALL, W.: The Structure of Magneto-Hydrodynamic Shock Waves. Proc. Roy. Soc. London A-233, 1956, pp. 367–376.

5. PAI, S. I.: On Exact Solutions of One Dimensional Flow Equations of Magneto-Gasdynamics. Proc. IX International Congress of Appl. Mech., 1956, pp. 17–25.

6. Pai, S. I.: One Dimensional Unsteady Flow of Magneto-Gasdynamics. Proc. Fifth Midwestern Conf. on Fluid Mech., Univ. of Michigan, 1957, pp. 251–261.

7. Pai, S. I.: Fluid Dynamics of Jets, Chap. II. D. Van Nostrand Co., Inc., Princeton, N. J., 1954.

8. Resler, E. L. Jr., and W. R. Sears: The Prospects for Magneto-Aerodynamics. Jour. Aero. Sci., vol. 25, No. 4, April 1958, pp. 235–245.

9. Resler, E. L. Jr., and W. R. Sears: Magneto-Gasdynamic Channel Flow. Jour. App. Math. and Phys. (ZAMP), vol. IXb, No. 5/6, 1958, pp. 509–518.

10. Rosa, R. J., and A. R. Kantrowitz: Magnetohydrodynamic Energy Conversion Techniques. AVCO Research Report 86, 1959.

11. Steg, L., and G. W. Sutton: Prospects of MHD Power Generation Astronautics, vol. 5, No. 8, 1959.

12. Way, S., et al.: Experiments with MHD Power Generation. ASME Publication 60-WA-328, 1960.

Chapter X

Wave Motion in a Plasma

1. Introduction. One of the most interesting problems of plasma dynamics is the wave motion of infinitesimal disturbances in a plasma. Such wave motion will bring out many characteristic features of plasma dynamics which is very interesting in itself and there are also many practical applications of the theory of wave motion (*5*) in many geophysical and astrophysical problems such as the theory of sunspots, etc.

One of the simplifying assumptions in the study of wave motion of infinitesimal amplitude is that the fundamental equations may be linearized. It is much easier to discuss the linearized equations than the original nonlinear equations. We shall derive the linearized equations of electromagnetodynamics in § 2.

The result of the linearized equation is not in itself very simple. There is still a great variety of oscillatory motions in plasma which may be very complex (*2–4*). We shall limit ourselves to the purely sinusoidal disturbances in a homogeneous medium and these disturbances are governed by the fundamental equations of electromagnetogasdynamics (5.1) except that we shall neglect the radiation phenomena and the non-electric force $\vec{F}_g$. Furthermore we shall consider the plasma as a single fluid (*9*) (cf. chapter XIII, § 5).

Before we discuss the general wave motion in plasma, we shall briefly review some of the simple results of wave motion due to an electromagnetic field only in § 3 and those of gasdynamic motions only in § 4. In § 5, we are going to discuss the wave motions under the magnetogasdynamic approximations while in § 6 we will discuss the electromagnetogasdynamic wave motion, in general, in which high frequency phenomena may occur so that the displacement current may not be neglected.

2. Linearized equations of electromagnetogasdynamics. We assume that originally the plasma is at rest with pressure p_0, temperature T_0 and density ρ_0. There is an external applied uniform magnetic field $\vec{H}_0 = \vec{i} H_x + \vec{j} H_y + \vec{k}\, O$ where $\vec{i}$, $\vec{j}$ and $\vec{k}$ are respectively the x-, y- and z-wise unit vector; and H_x and H_y are constant. There is no electric current, nor excess electric charge, nor external applied electric field. The plasma is perturbed by a small disturbance so that in the resultant disturbed motion, we have

$$u = u\,(x, t),\ v = v\,(x, t),\ w = w\,(x, t),\ p = p_0 + p'\,(x, t),\ T = T_0 + T'\,(x, t),$$

$$\rho = \rho_0 + \rho'\,(x, t),\ \vec{E} = \vec{E}\,(x, t),\ \vec{H} = \vec{H}_0 + \vec{h}\,(x, t),\ \vec{J} = \vec{J}\,(x, t),\ \rho_e = \rho_e\,(x, t) \tag{10.1}$$

where the perturbed quantities u, v, w, p', T', ρ', $\vec{E} = \vec{i} E_x + \vec{j} E_y + \vec{k} E_z$,

$\vec{h} = \vec{i}\, h_x + \vec{j}\, h_y + \vec{k}\, h_z$, $\vec{J} = \vec{i}\, J_x + \vec{j}\, J_y + \vec{k}\, J_z$ and ρ_e are assumed to be small and the second and higher order terms of these quantities are negligible. For simplicity, we assume that these perturbed quantities are functions of x and t only. Thus we will discuss only the wave propagation in the direction of x-axis. It is a straight forward process to generalize our results for the three dimensional case in which the perturbed quantities are functions of x, y, z and t.

Substituting equation (10.1) into the fundamental equations of plasma dynamics (4.1) to (4.8) and neglecting the higher order terms, we have the linear differential equations for the perturbed quantities as follows:

(a) Maxwell's equations (4.5) and (4.6) give

$$J_x + \frac{\partial \varepsilon E_x}{\partial t} = 0 \tag{10.2}$$

$$J_y + \frac{\partial \varepsilon E_y}{\partial t} = -\frac{\partial h_z}{\partial x} \tag{10.3}$$

$$J_z + \frac{\partial \varepsilon E_z}{\partial t} = +\frac{\partial h_y}{\partial x} \tag{10.4}$$

$$\frac{\partial \mu_e h_x}{\partial t} = 0 \tag{10.5}$$

$$\frac{\partial \mu_e h_y}{\partial t} = \frac{\partial E_z}{\partial x} \tag{10.6}$$

$$\frac{\partial \mu_e h_z}{\partial t} = -\frac{\partial E_y}{\partial x} \tag{10.7}$$

(b) The generalized Ohm's law (4.8) gives

$$J_x = \sigma (E_x - \mu_e w H_y) \tag{10.8}$$

$$J_y = \sigma (E_y + \mu_e w H_x) \tag{10.9}$$

$$J_z = \sigma (E_z + \mu_e u H_y - \mu_e v H_x) \tag{10.10}$$

(c) The equation of conversation of electric charge (4.7) gives

$$\frac{\partial \rho_e}{\partial t} + \frac{\partial J_x}{\partial x} = 0 \tag{10.11}$$

(d) The equation of state (4.1) gives

$$\frac{p'}{p_0} = \frac{\rho'}{\rho_0} + \frac{T'}{T_0} \tag{10.12}$$

(e) The equation of continuity (4.2) gives

$$\frac{\partial \rho'}{\partial t} + \rho_0 \frac{\partial u}{\partial x} = 0 \tag{10.13}$$

(f) The equation of motion (4.3) gives

$$\rho_0 \frac{\partial u}{\partial t} = -\frac{\partial p}{\partial x} + \tfrac{4}{3}\mu \frac{\partial^2 u}{\partial x^2} - \mu_e J_z H_y \tag{10.14}$$

$$\rho_0 \frac{\partial v}{\partial t} = \mu \frac{\partial^2 v}{\partial x^2} + \mu_e J_z H_x \tag{10.15}$$

$$\rho_0 \frac{\partial w}{\partial t} = \mu \frac{\partial^2 w}{\partial x^2} + \mu_e (J_x H_y - H_x J_y) \tag{10.16}$$

(g) The energy equation (4.4) gives

$$\rho_0 C_p \frac{\partial T'}{\partial t} = \frac{\partial p'}{\partial t} + \varkappa \frac{\partial^2 T'}{\partial x^2} \tag{10.17}$$

From equations (10.2) to (10.17), we see that the equations may be divided into three groups.

(i) The quantity h_x is given by equation (10.5) alone and independent of all the other equations. Furthermore, because of equation (4.9), we conclude that h_x = constant, which may be put equal to zero.

(ii) The second group consists of the quantities w, h_z, J_x, J_y, E_x, E_y, and ρ_e which are governed by equations (10.2), (10.3), (10.7), (10.8), (10.9), (10.11), and (10.16). Since it deals with the velocity component w and the magnetic field h_z which are perpendicular to the applied external magnetic field H_0, we may call this wave a transverse wave.

(iii) The third group consists of the quantities u, v, p', ρ', T', h_y, J_z, and E_z which are governed by equations (10.4), (10.6), (10.10), (10.12), (10.13), (10.14), (10.15), and (10.17). This wave is sometimes called the longitudinal wave.

In the following sections, we are looking for periodic solutions in which all the perturbed quantities are proportional to

$$e^{i\omega t - i\lambda x} \tag{10.18}$$

where ω is a given real quantities and $i = \sqrt{-1}$. Substituting these variables into the differential equations of these perturbed quantities (10.2) to (10.17), we obtain one determinantal equation for each group of these quantities. The eigenvalues λ of these determinantal equations give the different modes of wave propagation through the plasma.

3. Electromagnetic waves. First we consider the electromagnetic waves in which all the gasdynamic variables are zero. We consider only the electromagnetic variables. For the transverse wave, we have

$$J_x = \sigma E_x, \quad J_x + \frac{\partial \varepsilon E_x}{\partial t} = 0, \quad \frac{\partial \rho_e}{\partial t} + \frac{\partial J_x}{\partial x} = 0 \tag{10.19}$$

and

$$J_y = \sigma E_y, \quad \frac{\partial \mu_e h_z}{\partial t} = -\frac{\partial E_y}{\partial x}, \quad J_y + \frac{\partial \varepsilon E_y}{\partial t} = -\frac{\partial h_z}{\partial x} \tag{10.20}$$

From equation (10.19) with the relation (10.18), we find that for J_x, E_x

and ρ_e, $i\,\omega = -\sigma/\varepsilon$. For finite electric conductivity, these waves are damped out, i. e.,

$$i\,\omega = -\sigma/\varepsilon \tag{10.21}$$

and

$$E_x = E_{x0}\, e^{-i\lambda x}\, e^{-(\sigma/\varepsilon)\,t} \tag{10.21a}$$

Similar expressions may be written for J_x and ρ_e.

From equation (10.20), we have

$$\frac{1}{c^2}\cdot\frac{\partial^2 E_y}{\partial t^2}+\frac{1}{\nu_H}\cdot\frac{\partial E_y}{\partial t}-\frac{\partial^2 E_y}{\partial x^2}=0 \tag{10.22}$$

This is the well-known telegraphic equation which has been extensively discussed in standard textbook of electromagnetic theory (*10*). The variables J_y and h_z satisfy the same telegraphic equation. For a non-conducting medium, equation (10.22) reduces to the simple wave equation with the speed of propagation equal to the velocity of light c. With finite electric conductivity and relation (10.18), we have the following relations between ω and λ:

$$\lambda^2-\frac{\omega^2}{c^2}+\frac{i\,\omega}{\nu_H}=0 \tag{10.23}$$

For a given value of ω, we may calculate λ. In this case, for a real value of ω, λ is complex. It is a damped wave in a dissipative medium. Both the speed of propagation and the rate of damping depends on the value of electric conductivity.

For the longitudinal wave, it is easy to show that all the variables h_y, J_z and E_z also satisfy equation (10.22). For the sinusoidal wave of (10.18), the relation (10.23) holds true for the longitudinal wave too.

It is of interest to see what kind of wave we may have under the magnetogasdynamic approximations in which the displacement current may be neglected. Equation (10.20) gives

$$\frac{\partial^2 E_y}{\partial x^2}-\frac{1}{\nu_H}\cdot\frac{\partial E_y}{\partial t}=0 \tag{10.24}$$

For the sinusoidal wave of (10.18), we have

$$\lambda^2 + i\,\mu_e\,\sigma\,\omega = 0 \quad\text{or}\quad \lambda^2 = -\frac{i\,\omega}{\nu_H} \tag{10.25}$$

It represents a damped magnetic wave.

It is easy to show that the longitudinal waves of h_y, J_z and E_z under the magnetogasdynamic approximations also satisfy both equations (10.24) and (10.25). It is one of the basic mode of waves in magnetogasdynamics as we shall discuss in § 5.

4. Gasdynamic waves. Next we consider the gasdynamic waves in which all the electromagnetic variables are zero. We consider only the gasdynamic variables. For the transverse wave, we have simply

$$\rho_0\frac{\partial w}{\partial t}=\mu\frac{\partial^2 w}{\partial x^2} \tag{10.26}$$

and a similar equation for v. With the relation (10.18), equation (10.26) gives

$$\nu \lambda^2 + i\,\omega = 0 \tag{10.27}$$

where $\nu = \mu/\rho_0$ = coefficient of kinematic viscosity of the medium. It is the well-known damped wave in a viscous fluid (*7, 8*). This wave is independent of the compressibility effect of the medium. It is also one of the basic modes of wave propagation in magnetogasdynamics as we shall discuss in § 5.

For the longitudinal wave according to our definition in § 2, we have

$$\frac{p'}{p_0} = \frac{\rho'}{\rho_0} + \frac{T'}{T_0} \tag{10.28a}$$

$$\frac{\partial \rho'}{\partial t} + \rho_0 \frac{\partial u}{\partial x} = 0 \tag{10.28b}$$

$$\rho_0 \frac{\partial u}{\partial t} = -\frac{\partial p}{\partial x} + \tfrac{4}{3}\mu \frac{\partial^2 u}{\partial x^2} \tag{10.28c}$$

$$\rho_0 C_p \frac{\partial T'}{\partial t} = \frac{\partial p'}{\partial t} + \varkappa \frac{\partial^2 T'}{\partial x^2} \tag{10.28d}$$

Substituting the perturbed quantities in the expression of (10.18) into equation (10.28), we obtain the determinantal equation for the eigenvalue of λ as follows [cf. equation (10.33)]:

$$\left[\varkappa\left(\frac{1}{\rho_0} + \frac{i\,4\,\omega\,\nu}{3\,p_0}\right)\lambda^4 - \left\{\frac{\omega^2 \varkappa}{p_0} + \frac{4\,\nu\,\omega^2}{3\,T_0(\gamma-1)} - i\,\omega\,C_p\right\}\lambda^2 - \frac{i\,\omega^3}{T_0(\gamma-1)}\right] = 0 \tag{10.29}$$

The equation (10.29) represents a sound wave in a viscous and heat-conducting fluid. If $\nu = \varkappa = 0$, equation (10.24) becomes

$$\lambda^2 = \frac{\omega^2}{T_0(\gamma-1)\,C_p} = \frac{\omega^2}{\gamma R T_0} = \frac{\omega^2}{a_0^2} \tag{10.30}$$

in which a_0 is the sound velocity in the medium at rest. Hence equation (10.30) represents the sound wave in an inviscid and non-heat-conducting fluid. In general, the two solutions of λ^2 of equation (10.29) represent the two differnet modes of sound wave propagation in a viscous and heat-conducting fluid. Hence we see that for ordinary gasdynamics we have three different modes of wave motion. This result has been discussed in standard textbooks of gasdynamics (*7, 8*).

5. Magnetogasdynamic waves. Now we consider magnetogasdynamic waves in which we consider both the electromagnetic and the gasdynamic variables, but with the magnetogasdynamic approximations so that the displacement current and the excess electric charge may be neglected. For the transverse waves, we consider only the variables w and h_z because $J_x = 0$; and J_y, E_x and E_y can be expressed in terms of w and h_z. The eigenvalue λ is governed by the following relation:

$$\nu\,\nu_H\,\lambda^4 + [V_x^2 + i\,(\nu + \nu_H)\,\omega]\,\lambda^2 - \omega^2 = 0 \tag{10.31}$$

where $V_x = \sqrt{\mu_e/\rho_0}\,H_x$ = x-component of the velocity of Alfvén's wave. The solution of λ of equation (10.31) gives two modes of magnetogasdynamic waves which are developed from three basic modes of waves: one is the magnetic wave of (10.25) and the second is the viscous wave of (10.27) while the third is the so-called Alfvén's wave which is a special wave of magnetogasdynamics. If $V_x = 0$, equation (10.31) gives the solutions of (10.25) and (10.27). If $\nu = \nu_H = 0$, equation (10.31) gives

$$\lambda^2 = \frac{\omega^2}{V_x^2} \quad \text{or} \quad \frac{\omega^2}{\lambda^2} = V_x^2 \tag{10.32}$$

This is known as Alfvén's wave which has a speed of propagation of V_x. It was ALFVÉN (*1*) who first showed that if there is a homogeneous magnetic field H_x in an incompressible and inviscid fluid of density ρ_0 and of infinite conductivity $\sigma = \infty$, the disturbances in this fluid will propagate as a wave in the direction of H_x with a speed of V_x. From our analysis, we see that the velocity V_x is still a characteristic velocity in magnetogasdynamic waves. If $V_x \neq 0$, the viscous and the magnetic waves are coupled. Equation (10.31) gives the two solutions of these coupled waves. Similar to the gasdynamic waves, these transverse waves are independent of the compressibility effect of the medium.

For the longitudinal waves, we consider the variables u, v, h_y, p', ρ', and T' because both J_z and E_z may be expressed in terms of h_y. The eigenvalue is governed by the following relation:

$$\begin{gathered}\left[\varkappa\left(\frac{1}{\rho_0} + \frac{i\,4\,\omega\,\nu}{3\,p_0}\right)\lambda^4 - \left\{\frac{\omega^2\varkappa}{p_0} + \frac{4\,\nu\,\omega^2}{3\,T_0\,(\gamma - 1)} - i\,\omega\,C_p\right\}\lambda^2 - \frac{i\,\omega^3}{T_0\,(\gamma - 1)}\right] \\ \left[(\nu_H\,\lambda^2 + i\,\omega)\cdot(\nu\,\lambda^2 + i\,\omega) + V_x^2\,\lambda^2\right] \\ - \lambda^2\,V_y^2\,(i\,\omega + \nu\,\lambda^2)\cdot\left[\frac{\omega^2}{T_0\,(\gamma - 1)} - \frac{i\,\omega\,\varkappa\,\lambda^2}{p_0}\right] = 0\end{gathered} \tag{10.33}$$

where $V_y = \sqrt{\mu_e/\rho_0}\, H_y$ = y-component of velocity of Alfvén's wave. The first square bracket of equation (10.33) represents sound waves in a viscous heat-conducting fluid as we have shown in equation (10.29). If there is no transverse magnetic field, i. e., $H_y = V_y = 0$, there is no coupling between the sound waves and the magnetic waves of equation (10.31). In this case, we have two modes of sound waves given by equation (10.29) and two modes of magnetogasdynamic waves given by equation (10.31). If $V_y \neq 0$, there are couplings between the three basic modes: sound waves, viscous waves and simple magnetic waves. We shall get four different coupled magnetogasdynamic waves.

The effect of H_y on the sound waves may be seen clearly by considering the simple case of $\nu = \nu_H = V_x = \varkappa = 0$. Equation (10.33) reduces to

$$a_e^2 = \frac{\omega^2}{\lambda^2} = a_0^2 + V_y^2 \tag{10.34}$$

where a_e is the effective sound speed in an ideal plasma with $H_y \neq 0$ as we have shown in chapter IX, § 8 [cf. equation (9.39)].

For the case of ideal plasma under a general magnetic field, we have $\nu = \nu_H = \varkappa = 0$, equation (10.33) reduces to

$$\left(a_0^2 - \frac{\omega^2}{\lambda^2}\right)\cdot\left(V_x^2 - \frac{\omega^2}{\lambda^2}\right) - \frac{\omega^2}{\lambda^2}\,V_y^2 = 0 \tag{10.35}$$

There are two speeds of propagation of small disturbances given by equation (10.35) which may be designated by c_{fast} and c_{slow} such that

$$c_{\text{slow}} \leqq a_0 \leqq c_{\text{fast}} \tag{10.36a}$$

and

$$c_{\text{slow}} \leqq V_x \leqq c_{\text{fast}} \tag{10.36b}$$

For this case equation (10.32) still holds true. For the case of ideal plasma under a general magnetic field, we have three different speeds of propagation V_x, c_{slow} and c_{fast}. We shall discuss this point further in chapter XII when we discuss the method of characteristics of three dimensional flow in an ideal plasma.

6. Electromagnetogasdynamic waves. In this section we are going to consider the complete set of equations (10.2) to (10.17) without any further approximations. The results in this section include all the results of the above three sections (*11*). We shall give the detailed calculations of this complete case and show how to reduce this general result to various special cases discussed above.

For the transverse waves, we may use equations (10.8) and (10.9) to eliminate J_x and J_y. Hence we consider only the five variables w, h_z, E_x, E_y, and ρ_e which are governed by the following five equations:

$$-\mu_e \sigma w H_y + \left(\sigma E_x + \frac{\partial \varepsilon E_x}{\partial t}\right) = 0 \tag{10.37a}$$

$$\mu_e \sigma w H_x + \frac{\partial h_z}{\partial x} + \left(\sigma E_y + \frac{\partial \varepsilon E_y}{\partial t}\right) = 0 \tag{10.37b}$$

$$\frac{\partial \mu_e h_z}{\partial t} + \frac{\partial E_y}{\partial x} = 0 \tag{10.37c}$$

$$-\frac{\partial \sigma w \mu_e H_y}{\partial x} + \frac{\partial \sigma E_x}{\partial x} + \frac{\partial \rho_e}{\partial t} = 0 \tag{10.37d}$$

$$\left(\frac{\partial w}{\partial t} - \nu \frac{\partial^2 w}{\partial x^2} + \mu_e \sigma V_y^2 w + \mu_e \sigma V_x^2 w\right) - \frac{\mu_e \sigma H_y}{\rho_0} E_x + \frac{\mu_e \sigma H_x}{\rho_0} E_y = 0 \tag{10.37e}$$

Substituting the expression (10.18) for these unknowns w, h_z, E_x, E_y, and ρ_e, we obtain the following determinantal equation:

$$\begin{vmatrix} w & h_z & E_x & E_y & \rho_e \\ -\sigma \mu_e H_y & 0 & \sigma + i\omega\varepsilon & 0 & 0 \\ \sigma \mu_e H_x & -i\lambda & 0 & \sigma + i\omega\varepsilon & 0 \\ 0 & i\omega\mu_e & 0 & -i\lambda & 0 \\ i\lambda\sigma\mu_e H_y & 0 & -i\lambda\sigma & 0 & i\omega \\ i\omega + \nu\lambda^2 + \sigma\mu_e(V_y^2 + V_x^2) & 0 & -\dfrac{\sigma\mu_e H_y}{\rho_0} & \dfrac{\sigma\mu_e H_x}{\rho_0} & 0 \end{vmatrix} = 0 \tag{10.38}$$

Expanding equation (10.38) gives

$$\left(\frac{i\,\omega}{\nu_H}-\frac{\omega^2}{c^2}\right)\cdot\left[(i\,\omega+\lambda^2\,\nu)\cdot\left(\lambda^2-\frac{\omega^2}{c^2}+\frac{i\,\omega}{\nu_H}\right)+\frac{V_x{}^2}{\nu_H}\left(\lambda^2-\frac{\omega^2}{c^2}\right)\right]-$$
$$-\frac{\omega^2}{c^2}\cdot\frac{V_y{}^2}{\nu_H}\left(\lambda^2-\frac{\omega^2}{c^2}+\frac{i\,\omega}{\nu_H}\right)=0 \tag{10.39}$$

Since equation (10.39) is a quadratic equation of λ^2, there are two different modes of these transverse waves. If there is no magnetic field, i. e., $V_x = V_y = 0$, these two transverse waves are the light wave of (10.23) and the viscous wave of (10.27). If there is a magnetic field, there are couplings between these two basic modes. It is interesting to notice that the factor $(i\,\omega/\nu_H - \omega^2/c^2)$ represents the damped electromagnetic wave of (10.21).

For magnetogasdynamic approximations, we put $c \to \infty$ and then equation (10.39) reduces to equation (10.31). Under magnetogasdynamic approximation the transverse wave is independent of V_y.

These general transverse electromagnetogasdynamic waves are also independent of the compressibility effect of the medium.

For the longitudinal waves, we consider the variables u, v, p', ρ', T', h_y, and E_z where J_z is eliminated by equation (10.10). The fundamental equations for these seven variables are

$$\sigma\,\mu_e\,H_y\,u-\sigma\,\mu_e\,H_x\,v+\left(\sigma\,E_z+\frac{\partial\,\varepsilon\,E_z}{\partial\,t}\right)-\frac{\partial\,h_y}{\partial\,x}=0 \tag{10.40a}$$

$$\frac{\partial\,E_z}{\partial\,x}-\frac{\partial\,\mu_e\,h_y}{\partial\,t}=0 \tag{10.40b}$$

$$\frac{p'}{p_0}-\frac{\rho'}{\rho_0}-\frac{T'}{T_0}=0 \tag{10.40c}$$

$$\rho_0\frac{\partial\,u}{\partial\,x}+\frac{\partial\,\rho'}{\partial\,t}=0 \tag{10.40d}$$

$$\left(\frac{\partial\,u}{\partial\,t}-\tfrac{4}{3}\,\nu\,\frac{\partial^2\,u}{\partial\,x^2}+\frac{\sigma}{\rho_0}\,\mu_e{}^2\,H^2\,u\right)-\frac{\sigma}{\rho_0}\,\mu_e{}^2\,H_x\,H_y\,v+\frac{1}{\rho_0}\cdot\frac{\partial\,p'}{\partial\,x}+\frac{\sigma}{\rho_0}\,\mu_e\,H_y\,E_z=0 \tag{10.40e}$$

$$-\frac{\sigma}{\rho_0}\,\mu_e{}^2\,H_x\,H_y\,u+\left(\frac{\partial\,v}{\partial\,t}-\nu\,\frac{\partial^2\,v}{\partial x^2}+\frac{\sigma}{\rho_0}\,\mu_e{}^2\,H_x{}^2\,v\right)-\frac{\sigma}{\rho_0}\,\mu_e\,H_x\,E_z=0 \tag{10.40f}$$

$$-\frac{\partial\,p'}{\partial\,t}+\left(\rho_0\,C_p\,\frac{\partial\,T'}{\partial\,t}-\varkappa\,\frac{\partial^2\,T'}{\partial\,x^2}\right)=0 \qquad [10.40g)$$

Substituting the expression (10.18) for these unknowns u, v, p', ρ', T', E_z, and h_y, we obtain the following determinantal equation for the eigenvalue λ:

$$
\begin{array}{ccccccc}
u & v & p' & \rho' & T' & E_z & h_y
\end{array}
$$

$$
\begin{vmatrix}
\sigma \mu_e H_y & -\sigma \mu_e H_x & 0 & 0 & 0 & \sigma + i\omega\varepsilon & i\lambda \\
0 & 0 & 0 & 0 & 0 & -i\lambda & -i\omega\mu_e \\
0 & 0 & \dfrac{1}{p_0} & -\dfrac{1}{\rho_0} & -\dfrac{1}{T_0} & 0 & 0 \\
-i\lambda\rho_0 & 0 & 0 & i\omega & 0 & 0 & 0 \\
i\omega + \frac{4}{3}\nu\lambda^2 + \dfrac{\sigma\mu_e^2}{\rho_0} H_y^2 & -\dfrac{\sigma\mu_e^2}{\rho_0} H_x H_y & -\dfrac{i\lambda}{\rho_0} & 0 & 0 & \dfrac{\sigma\mu_e}{\rho_0} H_y & 0 \\
-\dfrac{\sigma\mu_e^2}{\rho_0} H_x H_y & i\omega + \nu\lambda^2 + \dfrac{\sigma\mu_e^2}{\rho_0} H_x^2 & 0 & 0 & 0 & -\dfrac{\sigma\mu_e}{\rho_0} H_x & 0 \\
0 & 0 & -i\omega & 0 & i\omega\rho_0 C_p + \varkappa\lambda^2 & 0 & 0
\end{vmatrix} = \tag{10.41}
$$

Expanding equation (10.41) gives

$$
\begin{aligned}
&\left[\varkappa\left(\frac{1}{\rho_0} + \frac{i\,4\,\nu\,\omega}{3\,p_0}\right)\lambda^4 - \left\{\frac{\omega^2\varkappa}{p_0} + \frac{4\,\nu\,\omega^2}{3\,T_0(\gamma-1)} - i\,\omega\,C_p\right\}\lambda^2 - \frac{i\,\omega^3}{T_0(\gamma-1)}\right] \\
&\left[\left(\lambda^2 + \frac{i\,\omega}{\nu_H} - \frac{\omega^2}{c^2}\right)\cdot(i\,\omega + \nu\lambda^2) + \frac{V_x^2}{\nu_H}\left(\lambda^2 - \frac{\omega^2}{c^2}\right)\right] \\
&-\left(\lambda^2 - \frac{\omega^2}{c^2}\right)\frac{V_y^2}{\nu_H}(i\,\omega + \nu\lambda^2)\cdot\left[\frac{\omega^2}{T_0(\gamma-1)} - \frac{i\,\omega\,\varkappa\,\lambda^2}{p_0}\right] = 0
\end{aligned}
\tag{10.42}
$$

The only difference of the general electromagnetogasdynamic waves given by equation (10.42) from those magnetogasdynamic waves given by equation (10.33) is in the expression of electromagnetic waves. If we put $c \to \infty$, equation (10.42) reduces to equation (10.33). If there is no transverse magnetic field, i. e., $H_y = V_y = 0$, there is no coupling between the sound wave, the electromagnetic wave, and the viscous wave. If $V_y \neq 0$, there are couplings between the three basic modes, sound waves, viscous waves and electromagnetic waves.

The effect of H_y on the sound speed in the general electromagnetogasdynamic wave may be obtained by putting $\nu = \nu_H = V_x = \varkappa = 0$, in equation (10.42) and we have

$$
\frac{\omega^2}{\lambda^2} = (a_0^2 + V_y^2)\,\frac{1}{\left(1 + \dfrac{V_y^2}{c^2}\right)} \tag{10.43}
$$

This speed is a little smaller than the corresponding effective sound speed of magnetogasdynamic waves given in equation (10.34).

References

1. Alfvén, H.: On the Existence of Electromagnetic Hydrodynamic Waves. Arkiv f. Math. Astro. Ock Fysik 29b, No. 2, pp. 1–7, 1943.
2. Banos, A. Jr.: Fundamental Wave Functions in an Unbounded Magneto-Hydrodynamic Field. I. General Theory. Phys. Rev., vol. 97, No. 6, pp. 1435–1443, 1955.
3. Banos, A. Jr.: Magneto-Hydrodynamic Waves in Incompressible and Compressible Fluids. Proc. Roy. Soc. London A-2333, pp. 350–366, 1955.
4. Banos, A. Jr.: Normal Modes Characterizing Magneto-Elastic Plane Waves. Phys. Rev., vol. 104, No. 2, pp. 300–305, 1956.

5. COWLING, T. G.: Magnetohydrodynamic. Chap. III, Interscience Publishers, Inc., N.Y., 1957.
6. LUDFORD, G. S. S.: Small Perturbation in Hydromagnetics, Free Oscillations in a Rectangular Box, And the Perfectly Conducting Gas. Tech. Note BN-143, Institute for Fluid Dynamics and Applied Mathematics, University of Maryland, July 1958.
7. PAI, S. I.: Viscous Flow Theory I—Laminar Flow. Chapter XIV, D. Van Nostrand Co., Inc., Princeton, N. J., 1956.
8. PAI, S. I.: Viscous Flow Theory II—Turbulent Flow. Chapter XIV, D. Van Nostrand Co., Inc., Princeton, N. J., 1957.
9. SPITZER, L. Jr.: Physics of Fully Ionized Gases. Chapter IV, Interscience Publishers, Inc., N.Y., 1956.
10. STRATTON, J. A.: Electromagnetic Theory. McGraw-Hill Book Co., Inc., N.Y., 1941.
11. VAN DE HULST, H. C.: Interstellar Polarization and Magneto-Hydrodynamic Waves. Problems of Cosmic Aerodynamics. Central Air Documents Office 1951, pp. 45–56.

Chapter XI

Shock Wave Propagation in Magnetogasdynamics

1. Introduction. In chapter IX § 4, we discussed the Rankine-Hugoniot relations for a normal shock in a transverse magnetic field. If there is no transverse magnetic field but if a longitudinal magnetic field exists the shock relations in magnetogasdynamics will be identical to that of ordinary gasdynamics. In this chapter, we shall first discuss the oblique shock relations in magnetogasdynamics (§ 2). In this case, we have to consider both the direction of the magnetic field and that of the flow velocity with respect to the shock front. In ordinary gasdynamics, it is well-known that by proper choice of the coordinate system it is always possible to reduce the oblique shock to a corresponding normal shock case. But in magnetogasdynamic shocks, it is in general not possible to reduce an oblique shock case to a simple normal shock case by proper choice of the coordinate system. However we shall show in § 2 that by proper choice of the coordinate system, the oblique shock wave relations may be simplified even though it can not be reduced to a simple normal shock. In § 3, we shall discuss the oblique shock structure in magnetogasdynamics.

In the rest of this chapter, we shall discuss a few simple flow problems involving shock waves. In § 4, we will discuss the shock wave propagation in an infinitely electrically conductive gas with transverse magnetic field and gravitation. In § 5, cylindrical shock waves produced by instantaneous energy release in magnetogasdynamics will be treated, particularly the case of similar solutions.

As we pointed out in chapter IV § 4, if the magnetic pressure number is small, the influence of the magnetic field on the flow of the plasma is negligible. In § 6, we shall discuss such a case in which a shock wave penetrates into a magnetic field. We are going to discuss the distortion of the magnetic dipole field as a consequence of the approach of the shock wave when the magnetic Reynolds number is large. This problem was first treated extensively by J. M. Burgers (*1*).

2. Oblique shock in magnetogasdynamics. Similar to the case of normal shock (chapter IX § 4) we are going to find the relations of the flow variables in magnetogasdynamics, i. e., $\vec{u}$, ρ, S and $\vec{H}$ of the uniform states in front of and behind a shock wave which may be considered as a surface of discontinuity. We shall assume that both the velocity vector $\vec{u}$ and the magnetic field $\vec{H}$ have arbitrary directions with respect to the shock front whose normal vector may be denoted by $\vec{n}$. The jump of any physical quantity Q across the shock front may be denoted by

$$[Q] = Q_1 - Q_0 \tag{11.1}$$

where subscript 1 refers to the value on the side to which the normal $\vec{n}$ points and subscript o refers to the value on the other side. The relation between various Q may be obtained from the differential equations of magnetogasdynamics as we did for the normal shock case in chapter IX § 4 or simply by using the conservation laws across the surface of discontinuity. These relations are as follows:

$$[H_n] = 0 \tag{11.2a}$$

$$[(u_n - U)\,\mu_e\,\vec{H} - \mu_e\,H_n\,\vec{u}] = 0 \tag{11.2b}$$

$$\left[(u_n - U)\,\rho\,\vec{u} + \left(p + \frac{\mu_e H^2}{2}\right)\vec{n} - \mu_e\,H_n\,\vec{H}\right] = 0 \tag{11.2c}$$

$$[(u_n - U)\,\rho] = 0 \tag{11.2d}$$

$$\left[(u_n - U)\cdot\left(\frac{\rho\,u^2}{2} + \rho\,U_m + \frac{\mu_e H^2}{2}\right) + u_n\left(p + \frac{\mu_e H^2}{2}\right) - H_n\,(\mu_e\,\vec{H}\cdot\vec{u})\right] = 0 \tag{11.2e}$$

where subscript n refers to the component of the vector in the direction of the normal $\vec{n}$. U is normal component of the velocity vector of the surface of discontinuity. In order that there existe such a surface of discontinuity, there must be a surface current along this surface of continuity with a value of

$$\vec{J^*} = \vec{n} \times [\vec{H}] \tag{11.3}$$

From equation (11.2b) we see that if we choose the coordinate system such that the flow velocity $\vec{u}$ is parallel to the magnetic field $\vec{H}$ on one side of the shock front and $(u_{n1} - U)\,\mu_e\,\vec{H}_1 - \mu_e\,H_n\,\vec{u}_1 = 0$ these two vectors $\vec{u}$ and $\vec{H}$ are automatically parallel on the other side of the shock front, and $(u_{n0} - U)\,\mu_e\,\vec{H}_0 - \mu_e\,H_n\,\vec{u}_0 = 0$.

If $H_n = 0$, equation (11.2) reduces to the set of equations (9.13) to (9.15) for the case of the shock wave in an transverse magnetic field. If $\vec{H} = \vec{n}\,H_n$, the shock relations for the gasdynamic variables are exactly the same as those in ordinary gasdynamics.

We shall follow FRIEDRICH's analysis (*3*) to give a quick survey of the possible types of shocks from equation (11.2). We first introduce the following notations

$$\tau = \rho^{-1} \tag{11.4}$$

and

$$\tilde{Q} = \tfrac{1}{2}\,(Q_0 + Q_1) \tag{11.5}$$

and

$$m = \rho\,(u_n - U)$$

where the normal $\vec{n}$ s so chosen that across a shock wave

$$(u_n - U) > 0 \tag{11.6}$$

If $u_n - U = 0$, the surface of discontinuity may be called a contact discontinuity.

From equation (11.2a), we have

$$H_{n0} = H_{n1} = H_n = \text{constant} \tag{11.7}$$

Equations (11.2b) to (11.2d) become

$$m\,\tilde{\tau}\,\mu_e\,[\vec{H}] + \mu_e\,\tilde{\vec{H}}\,[u_n] - H_n\,[\vec{u}] = 0 \tag{11.8a}$$

$$m\,[\vec{u}] + \vec{n}\,[p] + \vec{n}\,\mu_e\,\tilde{\vec{H}}\,.\,[\vec{H}] - \mu_e\,H_n\,[\vec{H}] = 0 \tag{11.8b}$$

$$m\,[\tau] - [u_n] = 0$$

The determinantal equation of the system of equation (11.8) which determines the eigenvalue of the flux m is

$$\tilde{\tau}^2\,m\,(\tilde{\tau}\,m^2 - \mu_e\,H_n^2)\,.\,\{\tilde{\tau}\,m^4 + (\tilde{\tau}\,[\tau]^{-1}\,[p] - \mu_e\,\tilde{H}^2)\,m^2 - [\tau]^{-1}\,[p]\,\mu_e\,H_n^2\} = 0 \tag{11.9}$$

There are four solutions for m from equation (11.9) which give the four different surfaces of discontinuity or shocks in magnetogasdynamics. A brief discussion of these shocks are given below:

(a) Fast and slow shocks. The last factor of equation (11.9) gives two solutions of m which satisfy the equation

$$\tilde{\tau}\,m^4 + (\tilde{\tau}\,[\tau]^{-1}\,[p] - \mu_e\,\tilde{H}^2)\,m^2 - [\tau]^{-1}\,[p]\,\mu_e\,H_n^2 = 0 \tag{11.10}$$

We may denote these solution by m_{fast} and m_{slow} which satisfy the following relations

$$m_{\text{slow}}^2 \leqq -[\tau]^{-1}\,[p] \leqq m_{\text{fast}}^2 \tag{11.11}$$

and

$$m_{\text{slow}}^2 \leqq \tilde{\tau}^{-1}\,\mu_e\,H_n^2 \leqq m_{\text{fast}}^2 \tag{11.12}$$

For a given value of m_{slow} or m_{fast}, the following jumps of the physical quantities across the shock front occur

$$[\vec{H}] = k_1\,m^2\,(\tilde{\vec{H}} - \vec{n}\,H_n) \tag{11.13}$$

$$[\vec{u}] = k_1\,m\,(\mu_e\,H_n\,\tilde{\vec{H}} - \tilde{\tau}\,m^2\,\vec{n}) \tag{11.14}$$

$$[\tau] = -\,k_1\,(\tilde{\tau}\,m^2 - \mu_e\,H_n^2) \tag{11.15}$$

where k_1 is an arbitrary constant.

It is interesting to see how the magnetic field jumps across these shocks. From equation (11.13) we have

$$[H^2] = -\,m^2\,[\tau]\,(\tilde{\tau}\,m^2 - \mu_e\,H_n^2)^{-1}\,(\tilde{H}^2 - H_n^2) \tag{11.16}$$

From equations (11.12) and (11.16), we see that the magnetic field strength H increases across a fast shock and decreases across a slow shock. Equation (11.13) also shows that the tangential component of the magnetic field jumps in its own direction.

FRIEDRICHS defined a complete switch-on shock by which a tangential component of the magnetic field may be produced when it was absence ahead of it and a switch-off skock by which a tangential component of the magnetic field may be wiped out if it was present ahead of it. FRIEDRICHS found that a complete switch-on shock is a fast shock whose strength must lie below a critical value and in front of which the Alfvén wave speed is supersonic; while a complete switch-off shock is a slow shock whose strength must exceed a critical value if the Alfvén's wave speed behind the shock is supersonic. A complete switch-off shock always exists if the Alfvén's wave speed behind the shock is subsonic.

It is interesting to notice that these fast and slow shocks correspond to the fast and slow speeds of propagation of small disturbances respectively which are given in equation (10.36). For a fast shock we have

$$u_n - U > c_{\text{fast}}, \text{ ahead of the shock on the side (0)}$$

$$V_x \leqq u_n - U < c_{\text{fast}}, \text{ behind the shock on the side (1)}$$

where x is the normal direction of the shock, i. e., $V_x = \sqrt{\mu_e/\rho} \cdot H_n$

For a slow shock, we have

$$u_n - U \leqq V_x, \text{ behind the shock on the side (1)}$$

$$V_x = u_n - U > c_{\text{slow}}, \text{ ahead of the shock on the side (0)}$$

(b) Transverse shock. It corresponds to the root

$$m_{tr} = (\tilde{\tau}^{-1} \mu_e H_n^2)^{\frac{1}{2}} \tag{11.17}$$

This shock corresponds to the transverse wave of equation (10.32). The jumps across this type of shock are

$$[\vec{H}] = k\, m\, \vec{\tilde{H}} \times \vec{n} \tag{11.18a}$$

$$[\vec{u}] = k\, \mu_e H_n\, \vec{\tilde{H}} \times \vec{n} \tag{11.18b}$$

$$[\tau] = 0 \tag{11.18c}$$

$$[S] = 0 \tag{11.18d}$$

where k is an arbitrary constant

From equation (11.18a) we have

$$[H^2] = 0 \tag{11.19}$$

Hence the strength of the magnetic field is unchanged across a transverse shock.

(c) Contact discontinuity. It corresponds to $m = 0$. If $H_n \neq 0$, the only jump for this case is $[\rho] \neq 0$ because $[S] \neq 0$. If $H_n = 0$, we can only conclude that

$$[p + \tfrac{1}{2} \mu_e H^2] = 0 \tag{11.20}$$

For practical problems, we may want to find the relations of the physical quantities Q_1 when all the quantities Q_0 and the shock speed are given. These relations have been discussed by LUST (*8*) and HELFNER (*7*). HELFNER also gave some numerical computations for various cases in reference (*7*). One may also use the above relations to calculate the attached stationary shock wave at the leading edge of a wedge in a compressible fluid of infinite electric conductivity (*2*). Cabannes found that the attached shock angle in many cases can exceed 90°.

3. Oblique shock wave structure in magnetogasdynamics. The oblique shock wave structure similar to the analysis given in Chapter IX, §§ 3—7 for a normal shock has been studied by Germain (*6*). He showed that for given mass, momentum and enelgy flux, there exist at most 4 states S_1, S_2, S_3, S_4 which may be connected through a shock, the value of the index being an increasing function of the entropy. A shock wave from a shock S_i to a state S_j may exist if $i < j$. These four solutions correspond to the singular points of the differential system in a similar manner as we have shown for the normal shock case. They are characterized by the value of the normal speed component relative to the three characteristic velocities of infinitesimal wave in an ideal plasma as shown in chapter X, § 5. Germain also showed that a fast shock (S_1 to S_2) is stable, i. e., is a limit of transition layer for arbitrary dissipative coefficients becoming vanishingly small. An "intermediate shock" (S_1 to S_3 or S_4; S_2 to S_3 or S_4) is never stable. A "slow shock" (S_3 to S_4) may be stable.

4. Shock wave propagation in an infinitely electrically conductive plasma with transverse magnetic field and gravitation (*9*). We are going to consider the one dimensional flow produced by a strong shock wave in a gas of infinite electrical conductivity subjected to a transverse magnetic filed and to gravity. The shock wave is produced by the motion of a body. We shall study only the flow outside the shock region, so that viscous stress and heat conduction may be neglected and the electrical conductivity may be considered as infinite; but the change of entropy must be taken into account. There are two ways to investigate this problem: (1) the Eulerian method and (2) the Lagrangian method. In the Eulerian method we may use the method of characteristics discusses in chapter IX, § 8. Since for strong shock, the entropy S and the ratio (H/ρ) are no longer constant in the whole flow field at all time, we have to use all three sets of characteristic curves. It is rather difficult to work with three sets of lines in a $x - t$ plane. We shall use the Lagrangian method in this section.

In the Lagrangian method, we let x be the coordinate of an element of volume of the plasma at time t. We write

$$x = \phi\,(s, t) \tag{11.21}$$

where s is a parameter, characterizing that particular element of volume of plasma and thus its path in the $x - t$ plane. In genelal the choice of s is immaterial, but in certain problems, a special choice of s will be more convenient than others. We will choose s such that $x = s$ so long as the element of volume was at rest.

By definition

$$u = \frac{d\,x}{d\,t} = \frac{\partial\,\phi}{\partial\,t} \tag{11.22}$$

The equation of motion is

$$\frac{d\,u}{d\,t} = \frac{\partial^2\,\phi}{\partial\,t^2} = -\frac{1}{\rho}\cdot\frac{\partial\,p}{\partial\,x} - \frac{1}{\rho}\cdot\frac{\partial}{\partial\,x}\left(\frac{\mu_e\,H^2}{2}\right) - g \tag{11.23}$$

where H is the magnitude of the magnetic field perpendicular to the flow direction x and g is the gravitational acceleration.

If we write

$$\Phi = \frac{\partial \phi}{\partial s} \tag{11.24}$$

equation (11.23) becomes

$$\Phi \rho \left(\frac{\partial^2 \phi}{\partial t^2} + g\right) = -\frac{\partial}{\partial s}(p + p_H) \tag{11.25}$$

where $p_H = (\mu_e H^2)/2$.

In the $x - t$ plane we shall draw the path of the shock wave as $x = X(t)$. Let ρ_0 be the density of the plasma at rest and ρ, the density of the plasma after crossing the shock wave. By the equation of continuity, we have

$$\rho \, \mathrm{d} x = \rho_0 \, \mathrm{d} x_0 \tag{11.26}$$

Now $\mathrm{d} x = \left(\frac{\partial \phi}{\partial s}\right)(\mathrm{d} s) = \Phi \, \mathrm{d} s$ and since $\mathrm{d} s = \mathrm{d} x_0$, we obtain

$$\rho \Phi = \rho_0 \tag{11.27}$$

where ρ_0 may or may not be a constant. In general, one must assume ρ_0 to be a function of s.

If we consider only the case where the flow behind the shock wave is continuous so that the entropy is a constant along each line of flow, the pressure may be expressed in the following form:

$$p = p_1 \left(\frac{\rho}{\rho_1}\right)^\gamma = p_1 \left(\frac{\Phi_1}{\Phi}\right)^\gamma \tag{11.28}$$

where the subscript 1 refers to the state immediately behind the shock front.

Similarly we have

$$H = H_1 \frac{\rho}{\rho_1} = H_1 \frac{\Phi_1}{\Phi} \tag{11.29}$$

Hence the equation of motion (11.23) becomes

$$\rho_0 \left(\frac{\partial^2 \phi}{\partial t^2} + g\right) = -\frac{\partial}{\partial s}\left[p_1 \left(\frac{\Phi_1}{\Phi}\right)^\gamma + p_{H1} \left(\frac{\Phi_1}{\Phi}\right)^2\right] \tag{11.30}$$

where $p_{H1} = \mu_e H_1^2/2$ = magnetic pressure immediately behind the shock front.

The velocity of the shock wave is

$$\xi = \frac{\mathrm{d} x}{\mathrm{d} t} \tag{11.31}$$

At the path of the shock we have

$$\phi(s, t) = s \tag{11.32}$$

hence, taking the time derivative along this path, we have

$$\Phi_1 = \frac{\xi - u_1}{\xi} \tag{11.33}$$

The boundary conditions for this problem are as follows:

When we consider the problem of a shock wave produced by the motion of a body (e. g. by a piston acting on a column of plasma in a tube), we take $s = 0$ for the element of volume of the plasma immediately adjacent to the body. If the motion of the body or piston is fully given, $\phi\,(0, t)$ will be completely known. In other cases, a condition or an equation may be prescribed, determining the motion of this body. In either case, this fixes a boundary condition for the function $\phi\,(s, t)$ referring to $s = 0$.

A second boundary condition refers to the path of the shock wave. It must be obtained from the conditions (11.32) together with (11.33) and the generalized Rankine-Hugoniot relations of magnetogasdynamics (chapter IX, § 4). The generalized Rankine-Hugoniot relations in the coordinate system in which the velocity in front of the shock wave is zero, are as follows:

$$\rho_0\,\xi = \rho_1\,(\xi - u_1)$$

$$\frac{p_1}{\rho_1\,\xi^2} + \left(\frac{\xi - u_1}{\xi}\right)^2 - \left(1 + \frac{p_0}{\rho_0\,\xi^2} + \frac{p_{H0}}{\rho_0\,\xi^2} - \frac{p_{H1}}{\rho_0\,\xi^2}\right)\frac{\xi - u_1}{\xi} = 0$$

$$H_0\,\xi = H_1\,(\xi - u_1) \tag{11.34}$$

$$\frac{p_1}{(\gamma - 1)\,\rho_1\,\xi^2} - \frac{1}{2}\left(\frac{\xi - u_1}{\xi}\right)^2 +$$

$$+\left(1 + \frac{p_0}{\rho_0\,\xi^2} + \frac{p_{H0}}{\rho_0\,\xi^2} - \frac{p_{H1}}{\rho_0\,\xi^2}\right)\frac{\xi - u_1}{\xi} - \frac{a_0^2}{(\gamma - 1)\,\xi^2} - \frac{1}{2} - 2\,\frac{p_{H0}}{\rho_0\,\xi^2}\left(1 - \frac{H_1}{H_0}\right) = 0$$

where subscript 0 refers to value immediately in front of the shock front and subscript 1, that immediately behind the shock front.

From equation (11.34) we have

$$u_1 = \frac{1}{\gamma + 1}\left\{\frac{\gamma + 3}{2}\,\xi - \frac{a_0^2}{\xi} - \frac{\gamma\,p_{H0}}{\rho_0\,\xi} - \right.$$
$$\left. - \frac{1}{2}\sqrt{\left[(\gamma - 1)\,\xi + \frac{2\,a_0^2}{\xi} + \frac{2\,\gamma\,p_{H0}}{\rho_0\,\xi}\right]^2 + 8\,(2 - \gamma)\cdot(\gamma + 1)\,\frac{p_{H0}}{\rho_0}}\right\} \tag{11.35}$$

Furthermore

$$p_1 = p_0 + \rho_0\,\xi\,u_1 + p_{H0}\,(1 - \Phi_1^{-2}) \tag{11.36}$$

and

$$H_1 = H_0\,\Phi_1^{-1} \tag{11.37}$$

For small values of s and t, a series development for $\phi\,(s, t)$ can be used success fully to find the solution $\phi\,(s, t)$ which satisfies the boundary conditions and the differential equations. For large values of s and t, a numerical method should be used to continue the solution of series development. A typical numerical example (*9*) in which the flow field behind the shock wave in an infinitely electrically conductive plasma of constant temperature and constant magnetic field subjected to gravity is produced by a piston motion at a constant velocity is given in Fig. 11.1. The corresponding case of ordinary gasdynamics is also shown in Fig. 11.1. The effect of the magnetic field is to increase the speed of the shock wave in comparison with the corresponding case without magnetic field. Hence for the same motion of the piston, the range of the flow field influenced by the shock wave with a magnetic field is larger than that in the corresponding case without a magnetic field.

5. Cylindrical shock waves produced by instantaneous energy release (*10*). We now consider an idealized problem in which a finite amount

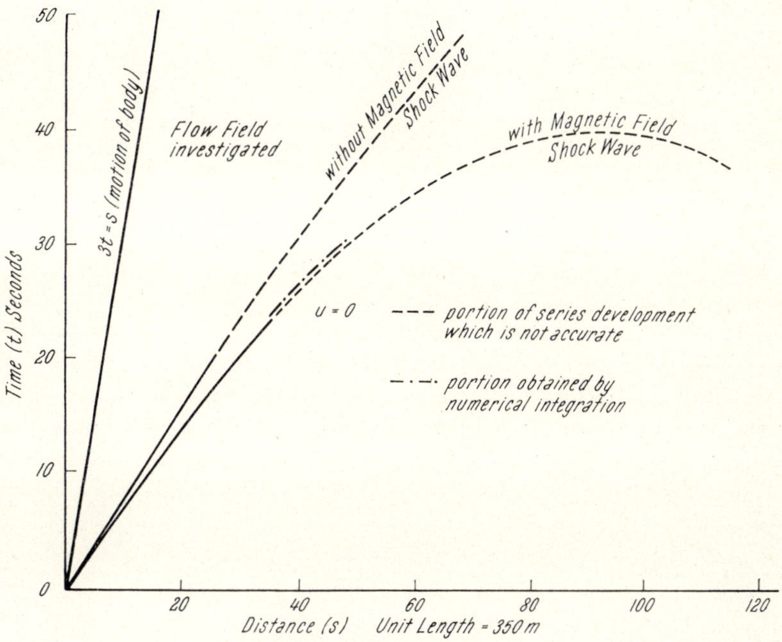

Fig. 11.1. Flow Field of a Shock Wave Propagated in a Gravitational Field with and without Transverse Magnetic Field.

$$\gamma = \frac{5}{3}, \quad \frac{P_{H\,0}}{p_{00}} = 0{,}1, \quad a_0 = 350\ \mathrm{m/sec.}, \quad g = 9{,}81\ \mathrm{m/sec}^2$$

of energy E_0 per unit length, in an infinitely concentrated form, is suddenly released along a straight line of infinite extent in a conducting plasma subjected to a tangential magnetic field. A cylindrical shock wave will be produced. In this problem, the velocity has only the radial component q_r while the magnetic field has only the tangential component H_θ Initially the plasma is assumed to be isothermal. The initial density distribution is assumed to be a function of the radial distance r, i. e.,

$$\rho_0 = \frac{B_\rho}{r^w} \tag{11.38}$$

and the initial magnetic field is of the form

$$H_{\theta 0} = H_0 = \frac{B_H}{r^{w_1}} \tag{11.39}$$

where B_ρ, B_H, w and w_1 are constant.

All the variables q_r, H_θ, p, ρ and T are functions of r and t only. We assume that $\mu = \varkappa = \nu_H = 0$. The fundamental equations for the five variables q_r, H_θ, p, ρ and T are as follows:

$$p = \rho R_p T \tag{11.40a}$$

$$\rho\left(\frac{\partial q_r}{\partial t} + q_r \frac{\partial q_r}{\partial r}\right) + \frac{\mu_e H_\theta^2}{r} = -\frac{\partial p}{\partial r} - \mu_e H_\theta \frac{\partial H_\theta}{\partial r} \tag{11.40b}$$

$$\frac{\partial \rho}{\partial t} + q_r \frac{\partial \rho}{\partial t} + \rho\left(\frac{\partial q_r}{\partial r} + \frac{q_r}{r}\right) = 0 \tag{11.40c}$$

$$\frac{\partial H_\theta}{\partial t} + q_r \frac{\partial H_\theta}{\partial r} + H_\theta\left(\frac{\partial q_r}{\partial r} + \frac{q_r}{r}\right) = 0 \tag{11.40d}$$

$$\left(\frac{\partial}{\partial t} + q_r \frac{\partial}{\partial r}\right) \cdot \left(\frac{p}{\rho^\gamma}\right) = 0 \tag{11.40e}$$

We shall solve the equations (11.40) for q_r, H_θ, p, ρ and T with proper boundary conditions at the shock wave.

First we are going to make some similarity considerations. There are three non-dimensional parameters γ, w and w_1 and six dimensional parameters r, t, E_0, B_ρ, B_H and μ_e in our problem. There are four fundamental units, i. e., length, mass, time and an electric unit. By dimensional analysis, we may form two more non-dimensional parameters from the six dimensional parameters. It is easy to show that these two new parameters may take the following form

$$\eta = \frac{r}{R} = \left(\frac{B_\rho}{\alpha E_0}\right)^{\frac{1}{4-w}} \frac{r}{t^{\frac{2}{4-w}}} \tag{11.41}$$

and

$$\tau = E_0 P_H^{n_1} B^{n_2} t^{n_3} \tag{11.42}$$

where α is a constant, $P_H = \mu_e B_H^2$ and

$$n_1 = \frac{4-w}{w-2(1+w_1)}, \quad n_3 = -2n_2 = \frac{4(1-w_1)}{w-2(1+w_1)} \tag{11.43}$$

If $w_1 = 1$, we have $n_1 = -1$, $n_2 = n_3 = 0$. Then τ is independent of r and t and it is a constant for a given problem. In this case, we have a single non-dimensional parameter η which depends on both r and t. We thus have the similar solutions, i. e., all the unknown may be expressed as function of η alone. Our partial differential equations may be transformed into total differential equations. To obtain similar solutions we write the unknowns in the following forms:

$$p = p_0 \frac{f_1(\eta)}{R^{m_1}} = \frac{B_p}{r^w} \cdot \frac{f_1(\eta)}{R^{m_1}}, \quad \rho = \rho_0 \psi(\eta) = \frac{B_p}{r^w} \psi(\eta)$$
$$q_r = \frac{\phi_1(\eta)}{R^n}, \quad H_\theta = H_{\theta 0} \frac{g_1(\eta)}{R^{m_2}} = \frac{B_H}{r} \cdot \frac{g_1(\eta)}{R^{m_2}} \tag{11.44}$$

If a similar solution exists the following relations must be fulfilled:

$$n = 1 - \tfrac{1}{2} w, \quad m_1 = 2n, \quad m_2 = 0 \tag{11.45}$$

and

$$R^n \frac{\mathrm{d} R}{\mathrm{d} t} = \text{constant} = A \tag{11.46}$$

Integration of equation (11.46) gives

$$R = \left(\frac{4-w}{2} A\right)^{\frac{2}{4-w}} t^{\frac{2}{4-w}} \tag{11.47}$$

Equation (11.47) agrees with the result of equation (11.41). R may be considered as the radius of the cylindrical shock wave.

We introduce the following notation

$$a^2 = \frac{\gamma B_p}{B_\rho} = \gamma R_\rho T_0 \tag{11.48}$$

and

$$V_H{}^2 = \frac{\mu_e B_H{}^2}{B_\rho} = \frac{p_H}{B_\rho} \tag{11.49}$$

The functions f_1, ϕ_1 and g_1 may be nondimensionalized as follows:

$$f_1(\eta) = \frac{A^2}{a^2} f(\eta), \quad \phi_1(\eta) = A \phi(\eta), \quad g_1(\eta) = \frac{A}{V_H} g(\eta) \tag{11.50}$$

The differential equations for f, ϕ, g and ψ as follows

$$f' = \frac{F}{F_1} \tag{11.51}$$

$$\text{where } F = \frac{w f}{\gamma \psi \eta} + \frac{\eta - \phi}{\gamma}\left[2n + \frac{(w-\gamma)}{\eta}\phi\right] + n\phi - \\ - \frac{g^2}{\gamma \eta^{2n} (\eta - \phi) \psi}\left[2n + \frac{(w-\gamma)}{\eta}\phi\right]$$

$$F_1 = \frac{1}{\psi \gamma f \eta^{2n}} \left[\eta^{2n} f - \psi \eta^{2n} (\eta - \phi)^2 + g^2\right]$$

$$\phi' = \frac{1}{\gamma f}\left[(\eta - \phi) f' + 2 n f - (\gamma - w)\frac{f \phi}{\eta}\right] \tag{11.52}$$

$$\psi' = \frac{\psi\left[\eta \phi' + \phi (1 - w)\right]}{\eta (\eta - \phi)} \tag{11.53}$$

$$g' = \frac{g \phi'}{\eta - \phi} \tag{11.54}$$

The functions f, ϕ, ψ and g can be obtained by integrating equations (11.51) to (11.54), starting from the values of f, ϕ, ψ and g given at the shock wave.

For a very strong shock wave, we have

$$f(1) = \frac{2\gamma}{\gamma+1}, \quad \psi(1) = \frac{\gamma+1}{\gamma-1}, \quad g(1) = \frac{V_H}{A}\left(\frac{\gamma+1}{\gamma-1}\right), \quad \phi(1) = \frac{2}{\gamma+1} \tag{11.55}$$

From the boundary conditions, we introduce two non-dimensional parameters γ and V_H/A into our problem, where V_H/A determines the strength of initial magnetic field with respect to the energy release. In reference (*10*), numerical examples are given for the case of $w = 0$ and two different values of V_H/A. It was found that the magnetic field has great influence on the pressure distributions but little influence on the density distribution. In general both the pressure and the velocity behind the shock wave are increased by the magnetic field.

6. The penetration of a shock wave into a magnetic field (*1*). In this section we are going to consider a problem in which the magnetic pressure number R_H is small and the magnetic Reynolds number is large. We may neglect the influence of the electromagnetic field on the flow field and consider only the influence of the flow field on the magnetic field. Particularly we shall investigate the problem of a plane shock wave moving in a plasma of high conductivity approaching a magnetic field. In such a problem it is convenient to use the vector and scalar potential of the electromagnetic field discussed in chapter III, § 5. We assume that the motion of the plasma is restricted in the x-direction. A plane shock wave of speed V is moving into a plasma at rest. Thus the velocity of the flow ahead of the shock wave is $u = 0$. We further assume that the velocity in the wake of the shock is a constant u. Under this condition, equation (3.38) becomes

$$\begin{aligned} \frac{\partial A_1}{\partial t} - \nu_H \nabla^2 A_1 &= -\frac{\partial \phi}{\partial x} \\ \frac{\partial A_2}{\partial t} + u\frac{\partial A_2}{\partial x} - v\frac{\partial A_1}{\partial y} - \nu_H \nabla^2 A_2 &= -\frac{\partial \phi}{\partial y} \\ \frac{\partial A_3}{\partial t} + u\frac{\partial A_3}{\partial x} - v\frac{\partial A_1}{\partial z} - \nu_H \nabla^2 A_3 &= -\frac{\partial \phi}{\partial z} \end{aligned} \tag{11.56}$$

where A_1, A_2 and A_3 are respectively the x-, y- and z-component of $\vec{A}$. We will attempt to obtain a solution for which $\phi = 0$ and for which A_1 has the same value as in the original field before the shock wave arrived. Since the velocity ahead of the shock is different from that behind the shock, we must solve the equation (11.56) separately for these two domains. However, it was found that A_1 is not affected by the motion of the plasma. Hence we shall consider only A_2 and A_3. The boundary conditions stating that A_2 and A_3, together with their derivatives with respect to x, must be continuous serve to link the solutions obtained in the two domains and determine the values of certain source functions which must be introduced along the shock wave path. The details of calculation has been given by BURGERS in reference (*1*). The main results are summaried as follows:

(a) Ahead of the shock wave, the magnetic induction $\vec{B}$ has the following approximate values:

$$B_1 = B_{10}(x, y, z)$$

$$B_2 = B_{20}(x, y, z)\left[1 + \frac{u}{V-u}\exp\frac{-V\eta}{\nu_H}\right] \quad (11.57)$$

$$B_3 = B_{30}(x, y, z)\left[1 + \frac{u}{V-u}\exp\frac{-V\eta}{\nu_H}\right]$$

where subscript 0 refers to the value of the original magnetic field. Terms of order of ν_H have been neglected in equation (11.57). The quantity η is the distance of the point x, y, z ahead of the shock front at the instant t, i. e., $\eta = x - V t$. Then ν_H is small, the influence of the shock wave ahead extends a short distance. When $\nu_H = 0$. the field ahead of the shock wave remains equal to the original field.

(b) Behind the shock wave, the magnetic induction $\vec{B}$ has the following values:

$$B_1 = B_{10}(\xi, y, z)$$

$$B_2 = B_{20}(\xi, y, z) \cdot \frac{d\xi}{dx} \quad (11.58)$$

$$B_3 = B_{30}(\xi, y, z) \cdot \frac{d\xi}{dx}$$

where ξ is the x-coordinate of the point of intersection of the streamline passing through the point of observation and that path of the shock wave. For a shock moving with a constant velocity V and having in its wake a uniform flow of velocity u, we have

$$\frac{d\xi}{dx} = \frac{V}{V-u} \quad (11.59)$$

Equation (11.58) shows that the magnetic field is transported from the point ξ, y, z with the motion of the plasma along the streamline, while at the same time the values of the components B_2 and B_3 are increased through multiplication with the "compression factor" $d\xi/dx$.

Burgers also considered the case where the conductivity is zero ahead of the shock wave while it can be treated as infinite behind the shock front in reference (*1*).

References

1. Burgers, J. M.: The Penetration of a Shock Wave into a Magnetic Field. Techn. Note BN-102, Institute for Fluid Dynamics and Applied Math. Univ. of Md., June 1957.

2. Cabannes, H.: Attached Stationary Shock Waves in Ionized Gases. Rev. Mod. Phys., Oct. 1960, pp. 973–976.

3. Hoffman, F. de, and E. Teller: Magnetohydrodynamic Shock. Phys. Rev. 80, 1950, pp. 692–703.

4. Friedrichs, K. O.: Nonlinear Wave Motion in Magneto-Hydrodynamics. Los Almos Laboratory Report, 1954.

5. Friedrichs, K. O., and H. Kranier: Nonlinear Wave Motion. Notes on Magnetohydrodynamics VIII, Institute of Math. Sci. New York Univ., 1958.

6. Germain, P.: Structure of Shock Waves in Magneto-Fluid Dynamics. Rev. Mod. Phys. Oct. 1960, pp. 951–958.

7. Helfer, H. L.: Magneto-Hydrodynamic Shock Waves. The Astro-Physical Jour. 117, 1953, pp. 177–199.

8. Lüst, R.: Magneto-hydrodynamische Stoßwellen in einem Plasma unendlicher Leitfähigkeit. Zeit. für Naturforschung 8a, Heft 5, 1953, pp. 277–284.

9. Pai, S. I.: Shock Wave Propagation in an Infinitely Electrically Conductive Gas with Transverse Magnetic Field and Gravitation. Techn. Note BN-109, Inst. for Fluid Dynamics and Applied Math., University of Maryland, 1957, Also ZAMM. Bd. 39, Heft 12, 1959, pp. 40–44

10. Pai, S. I.: Cylindrical Shock Waves Produced by Instantaneous Energy Release in Magneto-Gasdynamics. Tech. Note BN-120, Inst. for Fluid Dynamics and Applied Math., University of Maryland, 1958. Also Proc. 4th Indian Congress of Theoretical and Applied Mech., Dec. 28–31, 1958, pp. 89–100.

Chapter XII

Two- and Three-Dimensional Flows in Magnetogasdynamics

1. Introduction. In the analysis of one-dimensional flow (chapter IX), we actually considered only the average values of the velocity, pressure, etc., over certain sections. In order to study the details of the flow patterns, we have to find the velocity, pressure, etc., at every point in space. Hence we have to study the three dimensional flow and consider all three components of the velocity and magnetic field together with the pressure, density and temperature of the plasma as functions of the three spatial coordinates x, y, and z and time t. In general the three dimensional flow problem is too complicated to be analyzed. We have to make some reasonable simplifications in order to discuss some flow problems which are good approximations to the actual conditions. It is well known that the effects of viscosity μ and heat conductivity $\varkappa$ are negligibly small for high Reynolds number flow except in the boundary region near the solid wall or some other transition regions where the velocity and the temperature gradients are large. Hence in a majority of flow problems, the plasma may be considered as an inviscid and non-heat-conducting fluid. We shall make such an assumption in our treatments in §§ 2 to 5. For a plasma, besides viscosity and heat conductivity, there is a third diffusion property which may be characterized by the magnetic viscosity ν_H (or by electrical conductivity σ). If we assume $\nu_H = 0$ in addition to $\mu = 0$ and $\varkappa = 0$, we have an ideal plasma (cf. chapter IV, § 5). The properties of an ideal plasma have some similarities to those of an inviscid and non-heat-conducting gas. Particularly the system of the fundamental equations for the three dimensional and unsteady flows of an ideal plasma is hyperbolic just as that for a similar flow of ordinary gasdynamics of an inviscid fluid. We may use the method of characteristics to solve such problems. In § 2, we shall discuss the three dimensional unsteady flows of an ideal plasma.

In ordinary gasdynamics, the two-dimensional steady flows have been extensively investigated. It would be interesting to see what modifications occur in the corresponding problems in magnetogasdynamics. In § 3, we are going to discuss the two-dimensional steady flows over thin bodies for fluids of large or infinite electric conductivity. We shall assume that there is an externally applied magnetic field and study the perturbed velocity and magnetic field. Here a new parameter, the Alfvén's number which is $1/\sqrt{R_H}$, has a significant influence on the flow pattern. When the Alfvén's number is less than unity, we have upstream-inclined waves which is entirely different from the waves in ordinary gasdynamics. This phenomenon is similar to the upstream wake case as we mentioned in chapter VI, § 7.

For practical problem, the electric conductivity of the fluid is usually low. In order to have sufficient magnetogasdynamic effect on the flow field, a large "externally" applied magnetic field should be used. Such a magnetic field may have its origin inside the body. In § 4, we shall study the flow over thin magnetized bodies in fluids of low electric conductivity.

Another interesting problem is the hypersonic flow over a magnetized blunt body. Behind the detached shock and near the stagnation point of the blunt body, the temperature is very high and the gas will be ionized. As a result, the electromagnetic force would be large enough to influence the flow field. We shall consider such a flow in § 5.

In § 6, we shall consider the boundary layer flows of magnetogasdynamics so that the effects of viscosity and heat conductivity may be discussed. Here we shall divide our problems into two groups. In one group we shall assume that the electrical conductivity is very small so that the induced electromagnetic fields are negligible. In the other group, we shall assume that the electrical conductivity is very large so that the induced electromagnetic fields are very large and that the boundary layer of the flow field and that of the magnetic field are of the same order of magnitude. The flow conditions of those two groups differ from each other considerably.

2. Three-dimensional unsteady flow of an ideal plasma. The fundamental differential equations of the three dimensional unsteady flow of an ideal plasma in magnetogasdynamics are

$$\frac{\partial \vec{H}}{\partial t} = \nabla \times (\vec{u} \times \vec{H}) \tag{12.1a}$$

$$\frac{\partial \vec{u}}{\partial t} + (\vec{u} \cdot \nabla)\, \vec{u} - \frac{\mu_e}{\rho} (\vec{H} \cdot \nabla)\, \vec{H} = -\frac{1}{\rho} \nabla \left(p + \frac{\mu_e H^2}{2}\right) \tag{12.1b}$$

$$\frac{\partial \rho}{\partial t} + \nabla \cdot (\rho\, \vec{u}) = 0 \tag{12.1c}$$

$$\frac{\partial S}{\partial t} + (\vec{u} \cdot \nabla)\, S = 0 \tag{12.1d}$$

where S is the entropy and the pressure p is a function of the density and entropy S. From equation (12.1a) we also have the relation

$$\nabla \cdot \vec{H} = 0 \tag{12.2}$$

The properties of the system of equations (12.1) have been studied extensively by FRIEDRICHS (*1*, *2*). We are going to give some of his results in this section. FRIEDRICHS showed that equation (12.1) belong to the special class of symmetric hyperbolic equations. Solutions of this system of differential equations may possess "small" discontinuities only on certain characteristic manifolds $\mathfrak{S} = \mathfrak{S}(t)$ in (x, y, z)-space. Let $\vec{n}$ be a unit normal vector at each point of the surface $\mathfrak{S}(t)$ which moves characteristically with a velocity c_{ch} in the normal direction at each of its points. We may write

$$c_{ch} = u_n \pm c \tag{12.3}$$

where $u_n = \vec{u} \cdot \vec{n}$ is the normal component of the flow velocity at this point. We choose $c \geqq 0$. The possible discontinuities $\delta \vec{H}$, $\delta \vec{u}$, $\delta \rho$, δS of the quantities $\vec{H}$, u, ρ and S on the surface $\mathfrak{S}(t)$ are governed by the following relations:

$$\pm c \, \delta \vec{H} + \vec{H} \, \delta u_n - H_n \, \delta \vec{u} = 0 \tag{12.4a}$$

$$\pm \rho c \, \delta \vec{u} + a^2 \vec{n} \, \delta \rho + \mu_e \vec{n} (\vec{H} \cdot \delta \vec{H}) - \mu_e H_n \, \delta \vec{H} = 0 \tag{12.4b}$$

$$\pm c \, \delta \rho + \rho \, \delta u_n = 0 \tag{12.4c}$$

$$\pm c \, \delta S = 0 \tag{12.4d}$$

where $a = (\partial p / \partial \rho)^{\frac{1}{2}}$ = the speed of sound of the plasma (15.5)

$H_n = \vec{H} \cdot \vec{n}$ = the normal component of the magnetic field

and $\delta u_n = \vec{n} \cdot \delta \vec{u}$.

The characteristic velocity $\pm c$ is dertermined by the following determinantal equation obtained from equation (12.4):

$$\det. (12.4) = \rho c^2 (\rho c^2 - \mu_e H_n^2) \cdot [\rho c^4 - (\rho a^2 + \mu_e H^2) c^2 + a^2 \mu_e H_n^2] = 0 \tag{12.6}$$

All the eight roots of this equation are real because the differential equations are hyperbolic. The main difference of equation (12.6) from the corresponding case of ordinary gasdynamics is that in ordinary gasdynamics there is only one value of c differing from zero, i. e., the ordinary speed of sound while in the present case there are three values of c which differ from zero. Hence in magneto-gasdynamics there are three modes of propagation in each direction. Moreover these speeds depend on the direction, i. e., on the angle between the direction of propagation and the direction of the magnetic field. This point has already been noticed in our discussion of the wave motion of infinitesimal disturbances in chapter X. Now we are going to find various relations associated with the different roots of c as follows:

(a) Fast and Slow Disturbance waves.

These waves are associated with the last factor of equation (12.6), i. e.,

$$\rho c^4 - (\rho a^2 + \mu_e H^2) c^2 + a^2 \mu_e H_n^2 = 0 \tag{12.7}$$

Let us denote the two roots of c of equation (12.7) by c_{fast} and c_{slow} (these notations were first introduced by FRIEDRICHS) such that $c_{\text{fast}} > c_{\text{slow}}$. It is easy to show that the following relations hold

$$c_{\text{slow}} \leqq a \leqq c_{\text{fast}} \tag{12.8}$$

and

$$c_{\text{slow}} \leqq V_n \leqq c_{\text{fast}} \tag{12.9}$$

where $V_n = (\mu_e H_n^2 / \rho)^{\frac{1}{2}}$ = the Alfvén's velocity with respect to H_n.

If $H_n = 0$, we have the one-dimensional results of chapter IX, § 8, i. e.,

$$c_{\text{fast}} = a_e \quad \text{and} \quad c_{\text{slow}} = 0 \tag{12.10}$$

Equations (12.4) associated with c_{fast} and c_{slow} become

$$\delta \vec{H} = k \rho c^2 (\vec{H} - H_n \vec{n}) \tag{12.11a}$$

$$\delta \vec{u} = \pm k c (\mu_e H_n \vec{H} - \rho c^2 \vec{n}) \tag{12.11b}$$

$$\delta \rho = k \rho (\rho c^2 - \mu_e H_n^2) \tag{12.11c}$$

$$\delta S = 0 \tag{12.11d}$$

where k is any number different from zəro.

(b) Transverse wave and Alfvén waves. These waves are associated with the root

$$c = V_n = (\mu_e H_n^2/\rho)^{\frac{1}{2}} \tag{12.12}$$

The wave corresponds to the transverse wave (10.32). The possible discontinuities associated with this wave, given by equation (12.4) are

$$\delta \vec{H} = \pm k \rho c\vec{H} \times \vec{n} \tag{12.13a}$$

$$\delta \vec{u} = k \mu_e H_n \vec{H} \times \vec{n} \tag{12.13b}$$

$$\delta \rho = \delta S = 0 \tag{12.13c}$$

(c) Contact disturbances. These are associated with the root $c = 0$. If $H_n \neq 0$, equations (12.4) become

$$\delta H = k \vec{n}, \quad \delta S = k_1 \qquad \delta u = 0, \quad \delta \rho = 0 \tag{12.14}$$

where k and k_1 are arbitrary constants. This gives a contact discontinuity in entropy. In ordinary gasdynamics, such a discontinuity is usually associated with a discontinuity in the tangential velocity component. It is interesting to notice that in magnetogasdynamics if $H_n \neq 0$, a contact discontinuity does not permit a discontinuity in the tangential component of the flow velocity.

3. Two-dimensional steady flow over a thin body in fluids of large or infinite magnetic Reynolds number. In this section we shall treat the two dimensional problems of a uniform stream passing over a thin body under a uniform externally applied magnetic field. We assume that the body is so thin that the first order small perturbation theory is sufficient to describe the flow field. We are going to derive the fundamental equations of this linearized theory based on the following assumptions:

(i) The plasma is assumed to be inviscid and non-heat-conducting gas of large electric conductivity.

(ii) The flow is two dimensional and steady. Hence $\dfrac{\partial}{\partial t} = \dfrac{\partial}{\partial z} = 0$.

(iii) The velocity vector has the following components:

$$u = U + u', \quad v = v', \quad w = 0$$

where U is the velocity of the undisturbed uniform flow and u' and v' are the perturbed x- and y-components of the velocity respectively. We assume $U \gg u', v'$.

(iv) The externally applied magnetic field has the following components. $H_x = H_{0x}$ = constant, $H_y = H_{0y}$ = constant, $H_z = 0$. Hence magnetic the field has the components:

$$H_x = H_{0x} + h_x, \quad H_y = H_{0y} + h_y, \quad H_z = 0.$$

We assume that $H_{0x}, H_{0y} \gg h_x, h_y$.

(v) The flow is isentropic. It is true for the first order perturbation theory for the present case (cf. § 4). Hence the pressure is a function of density only, i. e., $p = p(\rho)$. We write

$$\rho = \rho_0 + \rho' \text{ and } \frac{dp}{d\rho} = a_0^2$$

where a_0 is the speed of sound of the undisturbed flow.

Substituting the above conditions into the fundamental equations of magnetogasdynamics (4.28) to (4.32), we obtain after some simplifications the following equation:

$$\begin{aligned}&[\beta^2 (1 - R_{Hx}) + M^2 R_{Hy}] \frac{\partial^4 \xi}{\partial x^4} + [1 - \beta^2 (R_{Hx} + R_{Hy})] \frac{\partial^4 \xi}{\partial x^2 \partial y^2} \\ &- R_{Hy} \frac{\partial^4 \xi}{\partial y^4} - 2 \sqrt{R_{Hx} R_{Hy}} \left(\frac{\partial^2}{\partial x \partial y}\right) \nabla^2 \xi = \frac{1}{R_\sigma} \left[\beta^2 \frac{\partial^2}{\partial x^2} + \frac{\partial^2}{\partial y^2}\right] \nabla^2 \frac{\partial \xi}{\partial x}\end{aligned} \qquad (12.15)$$

where $\xi = \partial h_y / \partial x - \partial h_x / \partial y$ = z-component of the electric current density (12.16)

$$\beta^2 = 1\text{-}M^2, \quad M = \frac{U}{a_0}, \quad R_{Hx} = \frac{\mu_e H_{0x}^2}{\rho_0 U^2}, \quad R_{Hy} = \frac{\mu_e H_{0y}^2}{\rho_0 U^2}, \quad R_\sigma = U L \sigma \mu_e$$

The parameters in our problem are the Mach number M of the uniform flow, the magnetic pressure numbers R_{Hx} and R_{Hy} and magnetic Reynolds number R_σ.

After ξ is determined, the other perturbed quantities can be obtained easily. For instance, the h_y is given by

$$\nabla^2 h_y = \left(\frac{\partial^2}{\partial x^2} + \frac{\partial^2}{\partial y^2}\right) h_y = \xi(x, y) \qquad (12.17)$$

We are going to examine equation (12.15) for various cases which will show the special features of the flow field in magnetogasdynamics. We shall consider the case of infinite magnetic Reynolds number $R_\sigma = \infty$, and the following orientations of the magnetic fields:

(i) Aligned fields. In this case, $H_{0y} = R_{Hy} = 0$, $R_\sigma = \infty$, equation (12.15) with the condition of vanishing disturbance far upstream becomes

$$\left(\beta^2 \frac{1 - R_{Hx}}{1 - \beta^2 R_{Hx}}\right) \frac{\partial^2 \xi}{\partial x^2} + \frac{\partial^2 \xi}{\partial y^2} = 0 \qquad (12.18)$$

Similar to the case of ordinary gasdynamics, equation (12.18) is a second order differential equation which may be either elliptic or hyperbolic depending on the sign of the first term. It is different from ordinary gasdynamics in that the Mach number M is no longer the only parameter which determines the property of the differential equation. Whether the equation is elliptic or hyperbolic, the combining effect of the Mach number and the magnetic pressure number is the determining factor. Fig. 12.1 shows the various regimes of the flow field.

The solution of the Laplace equation or the simple wave equation of (12.18) have been extensively discussed in standard textbook of aerodynamics. We are not going to give any example here.

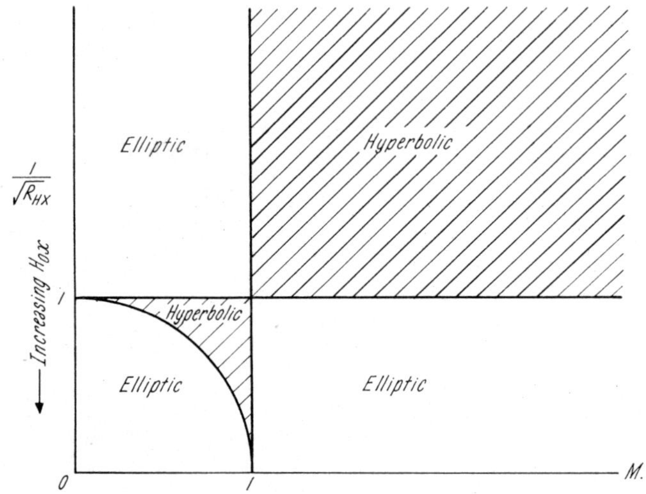

Fig. 12.1. Behavior of the Flow Fields for the Aligned Magnetic Fields.

(ii) Cross fields. In this case, $H_{0x} = R_{Hx} = 0$, $R_\sigma = \infty$, equation (12.15) becomes

$$\left[\frac{\partial^2}{\partial x^2} - R^2 \frac{\partial}{\partial y^2}\right] \times \left[B^2 \frac{\partial^2}{\partial x^2} + S^2 \frac{\partial^2}{\partial y^2}\right] \xi = 0 \tag{12.19}$$

$$\text{where } R^2 = \frac{R_{Hy}\beta^2 - 1 + \sqrt{4 M^2 R_{Hy}^2 + (1 + \beta^2 R_{Hy})^2}}{2(\beta^2 + M^2 R_{Hy})} \gtreqless 0$$

$$S^2 = \frac{R_{Hy}}{R^2} \gtreqless 0$$

$$B^2 = 1 - M^2 (1 - R_{Hy})$$

Equation (12.19) is different from the corresponding equation of ordinary gasdynamics. It is a fourth order differential equation. However if $R_{Hy} = 0$, equation (12.19) reduces to the corresponding equation of ordinary gasdynamics. The first operator of equation (12.19) is always a hyperbolic operator while the second operator may be an elliptic or a hyperbolic operator depending on the sign of B^2 because S^2 is always positive for non-vanishing M and R_{Hy}. We may treat this problem separately according to the nature of the second operator as follows:

(a) The hyper-liptic case. If $B^2 > 0$, the second operator is an elliptic operator. The solution of ξ may be written as follows:

$$\xi = \phi(x, y) + \psi(x, y) \tag{12.20}$$

where $\phi(x, y)$ is the solution of the homogeneous equation

$$\nabla_1^2 \xi = \left[B^2 \frac{\partial^2}{\partial x^2} + S^2 \frac{\partial^2}{\partial y^2} \right] \xi = 0 \tag{12.21}$$

and the function $\psi(x, y)$ is the particular solution of the following equation

$$\nabla_1^2 \psi = F(x - R^{-1} y) + G(x + R^{-1} y) \tag{12.22}$$

where F and G are arbitrary functions. The arbitrary functions F and G and the harmonic function ϕ are determined by the boundary conditions of a given problem (5).

(b) The purely hyperbolic regime. If $B^2 < 0$, both operators of equation (12.19) are hyperbolic. The general solution of ξ is simply

$$\xi = F_1(x - R^{-1} y) + G_1(x + R^{-1} y) + F_2(x - R_1^{-1} y) + G_2(x + R_1^{-1} y) \tag{12.23}$$

where F_1 etc. are arbitray functions to be determined from the boundary conditions (5), and $R_1^2 = -S^2/B^2$.

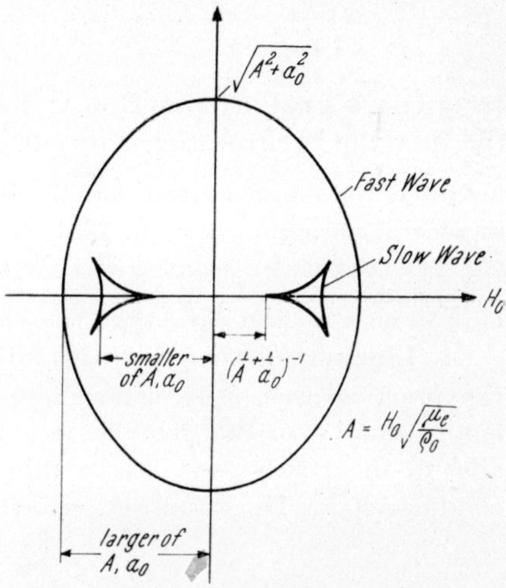

Fig. 12.2. Friedrichs Diagram

(iii) Arbitrary magnetic field orientation. In this case both H_{0x} and H_{0y} are different from zero. For $R_\sigma = \infty$, we should study equation (12.15) with the right-hand side terms to be zero. The resultant equation is again a fourth order differential equation. Since we consider the two dimensional problem, the flow pattern consists only the fast and slow waves discussed in § 2, i. e., equation (12.7). The wave patterns can be clearly shown by the well-know "Friedrichs diagram," i. e., Fig. 12.2, in which the abscissa is in the direction of the resultant magnetic field which is in the direction of $\vec{H}_0$ in our case. Fig. 12.2 shows the self-perpetuating shape of a disturbance produced at a point in an ideal plasma otherwise at rest. The standing magnetogasdynamic waves are the envelopes of the fast or slow waves of the Friedrichs disturbances produced by a moving body. These waves can be inclined "upstream" if the body's speed is such that an upstream envelope is appropriate. For instance, foı the case of aligned magnetic field, when R_{Hx} is larger than unity, upstream inclined waves are possible. Usually we classify the waves as

(a) Sub-Alfén's wave, when $R_H > 1$ and

(b) Super-Alfvén's wave, when $R_H < 1$.

New magnetogasdynamic phenomena occur at $R_H > 1$. Similaı case of upstream wake would occur for $R_H > 1$ for the viscous flow.

For finite electric conductivity, we should study equation (12.15). If the magnetic Reynolds number R_σ is very large, we will have phenomena similar to ordinary boundary layer flow, i. e., near the boundary of the body there will be a narrow region in which the influence of finite electric conductivity will be important. Outside the boundary layer, the fluid may be considered as an ideal plasma.

Another point of interest is that for isentropic flow in magnetogasdynamics the flow is not necessarily irrotational. The variation of vorticity $\vec{\omega} = \nabla \times \vec{u}$ in an ideal plasma is given by the following relation:

$$\frac{D\vec{\omega}}{Dt} = \left(\frac{\partial}{\partial t} + \vec{u} \cdot \nabla\right)\vec{\omega} = (\vec{\omega} \cdot \nabla)\vec{u} - \vec{\omega}(\nabla \cdot \vec{u}) + \nabla T \times \nabla S - \nabla \times \left[\frac{1}{\rho}\vec{H} \times (\nabla \times \vec{H})\right] \tag{12.24}$$

In ordinary gasdynamics $\vec{H} = 0$, for isentropic flow, if at one instance $\vec{\omega} = 0$, the flow will be irrotational for all subsequent times. However this result does not hold for magnetogasdynamics because the term $\nabla \times [1/\rho\, \vec{H} \times (\nabla \times \vec{H})]$ is in general not equal to zero. Only in special cases, the magnetic term in equation (12.24) vanish, for isentropic flow we have irrotational flow if at one instance $\vec{\omega} = 0$ such as the case that the flow starts from rest.

4. Linearized theory of airfoils in fluids of low electric conductivity. For practical cases, the electric conductivity of the plasma is small so that R_σ is also small. In this section, we consider a uniform stream of velocity U parallel to the x-axis past over a thin magnetized body in a fluid of low electric conductivity. The resultant velocity vector of the flow field is

$$\vec{q} = i\,U + \vec{q}' = i\,(U + u) + j\,v + k\,w \tag{12.25}$$

where i, j and k are respectively the unit vectors in the direction of the x-, y- and z-axis. The magnitude of the perturbed velocity $\vec{q}'$ is much smaller than U. The magnetic field strength is

$$\vec{H} = \vec{H}_0 + \vec{h} = i\,(H_{0x} + h_x) + j\,(H_{0y} + h_y) + k\,(H_{0z} + h_z) \tag{12.26}$$

where $\vec{H}_0$ is the original magnetic field of the thin body when there is no flow and the perturbed magnetic field $|\vec{h}|$ is much smaller than $|\vec{H}_0|$.

The current density is

$$\vec{J} = \nabla \times \vec{H} = \nabla \times \vec{h} = \sigma\,(\vec{E} + \vec{q} \times \vec{B}) \simeq \sigma\,(\vec{E}_0 + U\,\bar{i} \times \vec{B}_0) = \sigma\,\vec{G}\ \text{(say)} \tag{12.27}$$

where $\vec{B} = \mu_e \vec{H}$ and subscript o refers to the value of undisturbed field.

The pressure p, density ρ and entropy S of the fluid in the flow field are respectively:

$$p = p_0 + p', \quad \rho = \rho_0 + \rho', \quad S = S_0 + S' \tag{12.27 a}$$

The unknowns in our problem are q', h, p', ρ', and S'.

Substituting the variables in the fundamental equations of magnetogasdynamics and neglecting the higher order terms, we shall have a system of linear differential equations. From this system of equations, we find that the flow is not isentropic and the entropy is

$$S' = \frac{\sigma}{\rho_0 T_0 U} \int_{-\infty}^{x} G^2 \, dx \tag{12.28}$$

where we take $S_0 = 0$.

The non-dimensional differential equations for the gasdynamic variables Q (u, v, w, or ρ') are of the following form:

$$\nabla_M^2 Q = \left[(1 - M^2) \frac{\partial^2}{\partial x^2} + \frac{\partial^2}{\partial y^2} + \frac{\partial^2}{\partial z^2} \right] Q = F_Q \tag{12.29}$$

where $M = U/a_0$ and F_Q is a known function of the electromagnetic fields of the body, i. e., $\vec{E}_0$ and $\vec{B}_0$.

The solution of equation (12.29) may be written as follows:

$$Q = Q_0 + Q_1 \tag{12.30}$$

where Q_0 is the solution of the homogeneous equation of (12.29) with $F_Q = 0$ and the corresponding boundary conditions on the body. Hence Q_0 is exactly the ordinary gasdynamic solution. Q_1 is a particular solution of equation (12.29) with homogeneous boundary conditions on the body. Since F_Q is proportional to $R_\sigma R_H$, Q_1 is also proportional to $R_\sigma R_H$. Hence the contribution of the gasdynamic variables due to the electromagnetic fields $\vec{E}_0$ and $\vec{B}_0$ is proportional to $R_\sigma R_H$. We may thus use the electromagnetic fields $\vec{E}_0$ and $\vec{B}_0$ as a control device to change the gasdynamic forces on the magnetized thin body. The effectiveness of this control device is proportional to $R_\sigma R_H$.

5. Hypersonic flow past a blunt body in magnetogasdynamics. When a uniform hypersonic flow passes over a magnetized blunt body, there will be a detached shock in front of the body. Behind the detached shock and near the stagnation point of the blunt body, the temperature is very high and the gas will be ionized. As a result, the electromagnetic forces would be large enough to influence the flow field. We shall discuss a simple case by making the following assumptions:

(i) The shock front is assumed to be spherical and a body shape is found to be consistent with the shock.

(ii) We consider the flow near the stagnation point only.

(iii) The free stream magnetic field is the one due to a dipole at the origin of the body.

(iv) In the free stream, the electric conductivity of the gas is zero while behind the shock the electric conductivity of the gas is a constant σ.

(v) The gas is assumed to be inviscid and non-heat conducting.

It is convenient to use the polar coordinates (r, θ, ϕ). Because the flow is axi-symmetric, all the variables are independent of ϕ.

The boundary conditions at the shock which radius is c are as follows:

(i) Since the electric conductivity of the gas in front of the shock is zero, the tangential and normal components of the magnetic field at the shock should be continuous. In the free stream, we assume that the magnetic induction is

$$\vec{B}_0 = (2\,\alpha\, c^3 \cos\theta/r^3)\, i_r + (\alpha\, c^3 \sin\theta/r^3)\, i_\theta \tag{12.31}$$

where α is a constant and i_r and i_θ are respectively the unit vector in the r and θ direction.

Immediately behind the shock we have

$$\vec{B}_s = 2\,\alpha \cos\theta\, i_r + \alpha \sin\theta\, i_\theta \tag{12.32}$$

(ii) For the gasdynamic variables, the ordinary Rankine-Hugoniot relations hold. Immediately behind the shock we have

$$\vec{q}_s = -\varepsilon\, U \cos\theta\, i_r + U \sin\theta\, i_\theta \tag{12.33}$$

and

$$p_s = p_0 + \rho_0\, U^2 \cos^2\theta \quad (1 - \varepsilon) \tag{12.34}$$

where $\varepsilon = \rho_0/\rho_s$ and subscript s refers to the value immediately behind the shock and subscript o refers to the value in the free stream.

Since we consider the flow near the stagnation point, we may use the approximation

$$\cos\theta \simeq 1 - \theta^2/2 \quad \text{and} \quad \sin\theta \simeq \theta \tag{12.35}$$

Under these approximations, the velocity vector behind the shock may be written as

$$\vec{q} = [2\, f(r)\,(1 - \theta^2/2)/r^2]\, i_r - [f'(r)\,\theta/r]\, i_\theta \tag{12.36}$$

where prime refers to differentation with respect to r. The corresponding magnetic field behind the shock may be written as

$$\vec{B} = [2\, m(r)\,(1 - \theta^2/2)/r^2]\, i_r - [m'(r)\quad \theta/r]\, i_\theta \tag{12.37}$$

If we introduce the non-dimensionale variables:

$$R = r/c \tag{12.38a}$$

$$F(R) = f(r)/U\, c^2 \tag{12.38b}$$

$$M(R) = m(r)/\alpha\, c^2 \tag{12.38c}$$

The differential equation for $F(R)$ and $M(R)$ are respectively

$$F\,(F''' - 2\,F''/R - 2\,F'/R^2 + 8\,F/R^3) = -(2/R^2)\,Q\,\varepsilon\, M\,(-F\,M'' + F''\,M + {} + 4\,F\,M'/R - 4\,F'\,M/R) \tag{12.39}$$

and

$$-(M'' - 2\,M/R^2) = (2/R^2)\,R_\sigma\,(-F\,M' + F'\,M) \tag{12.40}$$

$$\text{where } R_\sigma = \sigma\,\mu_e\, c\, U \text{ and } Q = \sigma\,\alpha^2\, c/\rho_0\, U = R_\sigma\, R_H \tag{12.41}$$

The boundary conditions for F and M are

$$M(1) = 1$$

$$M'(1) = -1$$

$$F(1) = -\varepsilon/2 \tag{12.42}$$

$$F'(1) = -1$$

$$F''(1) = -(1/\varepsilon) + 2(1-\varepsilon) - 4Q(1+\varepsilon/2)$$

BUSH (*1*) integrates numerically equations (12.39) and (12.40) with the boundary conditions (12.41). He found that the detached shock distance increases with Q and that the velocity gradient at the body decreases as Q increases and that the pressure relief on the body increases with Q. BUSH studied only two values of R_σ, i. e., 1/5 and 1. He found that the effects of R are negligibly small in his results.

6. Two-dimensional boundary layer flow. It is well known that for the flow problems involving Reynolds numbers, the effects of viscosity and heat conductivity are important only in the boundary layer region near the solid body. The same condition is true in magnetogasdynamics. In this section we are going to investigate the two dimensional flow of a plasma over a flat plate. Let the plate lie in the x-z-plane and the flow of the free stream be in the x-direction. There is an externally applied magnetic field in the y-direction of the plate. Since the magnetic field depends greatly on the magnetic Reynolds number, we treat the cases of very small magnetic Reynolds number and very large magnetic Reynolds number separately.

(a) Case of very small magnetic Reynolds number. In this case the induced magnetic field is negligible. We may use the given value of the external applied magnetic field H_{y0} as the total magnetic field, i. e., $\vec{H} = \vec{j}\, H_{y0}$, where $\vec{j}$ is the unit vector in the y-direction. We assume that there is no external applied electric field. The current density $\vec{J}$ has a component in the z-direction only, i. e.,

$$\vec{J} = \vec{k}\, \sigma\, u\, \mu_e\, H_{y0} \tag{12.43}$$

where $\vec{k}$ is the unit vector in the z-direction. The ponderomotive force has a component in the x-direction only, i. e.,

$$\vec{F}_e = \mu_e\, \vec{J} \times \vec{H} = -\vec{i}\, \sigma\, u\, \mu_e^2\, H_{y0}^2 \tag{12.44}$$

where $\vec{i}$ is the unit vector in the x-direction. The Joule heat in the present case is

$$\frac{J^2}{\sigma} = \sigma\, u^2\, \mu_e^2\, H_{y0}^2 \tag{12.45}$$

The terms of viscous forces and of heat conduction may be simplified by the boundary layer approximation of ordinary gasdynamics. Finally we have the following boundary layer equations for the two-dimensional flow of a plasma over a flat plate:

$$\frac{\partial u}{\partial t} + u\frac{\partial u}{\partial x} + v\frac{\partial u}{\partial y} = -\frac{1}{\rho}\cdot\frac{\partial p}{\partial x} + \frac{1}{\rho}\cdot\frac{\partial}{\partial y}\left(\mu\frac{\partial u}{\partial y}\right) - \frac{\sigma u}{\rho}\,\mu_e^2\, H_{y0}^2 \tag{12.46a}$$

$$\frac{\partial p}{\partial y} = 0 \tag{12.46b}$$

$$\frac{\partial \rho}{\partial t} + \frac{\partial \rho u}{\partial x} + \frac{\partial \rho v}{\partial y} = 0 \tag{12.46c}$$

$$\rho\left(\frac{\partial C_p T}{\partial t} + u\frac{\partial C_p T}{\partial x} + v\frac{\partial C_p T}{\partial y}\right) = $$

$$= \frac{\partial p}{\partial t} + u\frac{\partial p}{\partial x} + \frac{\partial}{\partial y}\left(\varkappa\frac{\partial T}{\partial y}\right) + \mu\left(\frac{\partial u}{\partial y}\right)^2 + \sigma u^2 \mu_e^2 H_{y0}^2 \tag{12.46d}$$

$$p = \rho R_p T \tag{12.46e}$$

Since equations (12.46) are essentially the same type of equations as those of ordinary boundary layer equations, the methods of solution of ordinary boundary layer equations may be applied to equations (12.46) too. For an incompressible fluid, equations (12.46) reduce to the equations of boundary layer flow in magnetohydrodynamics discussed in chapter VI § 6.

For equations (12.46a) and (12.46d) we have for the steady flow case: the following relations:

$$\rho\left(u\frac{\partial h_0}{\partial x} + v\frac{\partial h_0}{\partial y}\right) = \frac{\partial}{\partial y}\left(\mu\frac{\partial h_0}{\partial y}\right) + \left(\frac{1}{P_r} - 1\right)\frac{\partial}{\partial y}\left(\mu\frac{\partial C_p T}{\partial y}\right) \tag{12.47}$$

If the Prandtl number $P_r = C_p \mu/\varkappa = 1$, we see that

$$h_0 = C_p T + \frac{u^2}{2} = \text{constant} \tag{12.48}$$

is a particular integral of equation (12.47). This is known as the Busemann relation (*6*) which holds in the present case too. However the Crocco relation of ordinary boundary layer flow does not hold here because of the term of the ponderomotive force in equation (12.46a).

(b) Case of very large magnetic Reynolds number (*4*). In this case, the induced magnetic field is not negligible. We shall assume that the external applied magnetic field is applied only on the plate. There is no external applied magnetic field in the front of the plate. Furthermore we shall assume that the magnetic Reynolds number is of the same order of magnitude as the ordinary Reynolds number. Hence we have boundary layer regions for velocity, temperature and magnetic field. Since the magnetic Reynolds number is of the same order of magnitude as the ordinary Reynolds number and Prandtl number is of the order of unity, these three boundary layer regions are of the same order of magnitude. Let δ be a measure of the thickness of this boundary layer. The characteristic length, λ, of the plate is much larger than δ. We are going to find the fundamental equations of the boundary layer flow for this case.

First we shall find the order of magnitude of u and v as well as the order of magnitude of H_x and H_y in the boundary layer. From the theory of ordinary boundary layer, we have

$$\frac{V}{U} \sim \frac{\delta}{\lambda} \ll 1 \tag{12.49}$$

where U and V are respectively the typical value of u and v in the boundary layer flow.

From equation (12.2) we have

$$\frac{\partial H_x}{\partial x} + \frac{\partial H_y}{\partial y} = 0 \tag{12.50}$$

We integrate equation (12.50) across the boundary layer with the boundary conditions: $y = 0$, $H_y = H_{y0}$; $y = \delta$, $H_y = 0$. We have

$$H_{y0} = \frac{H_{x0}\,\delta}{\lambda}\left[\int_0^1 \frac{\partial \bar{H}_x}{\partial \bar{x}}\,\mathrm{d}\bar{y}\right] \tag{12.51}$$

where $\bar{H}_x = H_x/H_{x0}$, $y = \bar{y}\,\delta$, $x = \lambda\,\bar{x}$, H_{x0} is the typical value of H_x in the boundary layer and H_{y0} may be considered as a typical value of H_y in the boundary layer. Since the non-dimensional term in the bracket is of the order of unity, we conclude that H_{y0}/H_{x0} is of the order of δ/λ, i. e., $(H_{y0}/H_{x0}) \ll 1$ in the boundary layer.

Now we put the fundamental equations of magnetogasdynamics in non-dimensional form by setting:

$$x = \lambda\,\bar{x},\ y = \delta\,\bar{y},\ u = \bar{u}\,U,\ v = \bar{v}\,V,\ H_x = H_{x0}\,\bar{H}_x,\ H_y = H_{y0}\,\bar{H}_y,\ t = \bar{t}\,\lambda/U,$$
$$\rho = \rho_0\,\bar{\rho} \tag{12.52}$$
$$T = T_0\,\bar{T},\ p = p_0\,\bar{p},\ \mu = \mu_0\,\bar{\mu},\ \varkappa = \varkappa_0\,\bar{\varkappa},\ Cp = C_{p0}\,\bar{C}_p,\ \nu_H = \nu_{H0}\,\bar{\nu}_H$$

where subscript o refers to some typical value and bar refers to the non-dimensional quantity which is of the order of unity.

The equations of the magnetic field in non-dimensional form are as follows:

$$\begin{array}{l}
\dfrac{\partial \bar{H}_x}{\partial \bar{t}} + \bar{u}\dfrac{\partial \bar{H}_x}{\partial \bar{x}} + \dfrac{V}{U}\cdot\dfrac{\lambda}{\delta}\bar{v}\dfrac{\partial \bar{H}_x}{\partial \bar{y}} + \bar{H}_x\dfrac{\lambda}{\delta}\cdot\dfrac{V}{U}\cdot\dfrac{\partial \bar{v}}{\partial \bar{y}} - \dfrac{H_{y0}}{H_{x0}}\dfrac{\lambda}{\delta}\bar{H}_y\dfrac{\partial \bar{u}}{\partial \bar{y}} \\
\quad 1 \qquad\quad 1 \qquad\quad \delta\quad \dfrac{1}{\delta} \qquad\qquad\qquad 1 \qquad\qquad\qquad \delta\quad \dfrac{1}{\delta} \\
\quad = \dfrac{1}{R_\sigma}\cdot\dfrac{\lambda}{\delta}\cdot\dfrac{\partial}{\partial \bar{y}}\left[\bar{\nu}_H\left(\dfrac{\lambda}{\delta}\cdot\dfrac{\partial \bar{H}_x}{\partial \bar{y}} - \dfrac{H_{y0}}{H_{x0}}\cdot\dfrac{\partial \bar{H}_y}{\partial \bar{x}}\right)\right] \\
\qquad \delta^2 \quad \dfrac{1}{\delta} \qquad\qquad\qquad \dfrac{1}{\delta} \qquad\qquad \delta
\end{array} \tag{12.53}$$

$$\begin{array}{l}
\dfrac{\partial \bar{H}_y}{\partial \bar{t}} + \bar{u}\dfrac{\partial \bar{H}_y}{\partial \bar{x}} + \dfrac{V}{U}\cdot\dfrac{\lambda}{\delta}\bar{v}\dfrac{\partial \bar{H}_y}{\partial \bar{y}} - \dfrac{H_{x0}}{H_{y0}}\cdot\dfrac{V}{U}\bar{H}_x\dfrac{\partial \bar{v}}{\partial \bar{x}} + \bar{H}_y\dfrac{\partial \bar{u}}{\partial \bar{x}} = \\
\quad 1 \qquad\quad 1 \qquad\quad \delta\quad \dfrac{1}{\delta} \qquad\quad \dfrac{1}{\delta} \qquad \delta \qquad\qquad 1 \\
\quad = \dfrac{1}{R_\sigma}\cdot\dfrac{\partial}{\partial \bar{x}}\left[\bar{\nu}_H\left(\dfrac{\partial \bar{H}_y}{\partial \bar{x}} - \dfrac{\lambda}{\delta}\cdot\dfrac{H_{x0}}{H_{y0}}\cdot\dfrac{\partial \bar{H}_x}{\partial \bar{y}}\right)\right] \\
\qquad \delta^2 \qquad\qquad\qquad 1 \qquad \dfrac{1}{\delta}\quad \dfrac{1}{\delta}
\end{array} \tag{12.54}$$

The order of magnitude of various terms is indicated under the equations. The same procedure is followed below. The magnetic Reynolds number $R_\sigma = U\,\lambda/\nu_{H\,0}$ should be taken as of the order of $1/\delta^2$ so that the diffusion term and the convection term of the magnetic field are of the same order of magnitude in the boundary layer.

The non-dimensional form of the equation of state is

$$\frac{\bar p}{\bar\rho\,\bar T} = \frac{R_p \rho_0 T_0}{p} \tag{12.55}$$

$$1 \qquad\qquad 1$$

The non-dimensional form of the equation of continuity is

$$\frac{\partial\bar\rho}{\partial\bar t} + \frac{\partial\bar\rho\,\bar u}{\partial\bar x} + \frac{V}{U}\cdot\frac{\lambda}{\delta}\cdot\frac{\partial\bar\rho\,\bar v}{\partial\bar y} = 0 \tag{12.56}$$

$$1 \qquad 1 \qquad \delta \quad \frac{1}{\delta}$$

The non-dimensional form of the equations of motion are

$$\frac{\partial\bar u}{\partial\bar t} + \bar u\frac{\partial\bar u}{\partial\bar x} + \bar v\frac{\partial\bar u}{\partial\bar y}\cdot\frac{\lambda}{\delta}\cdot\frac{V}{U} - R_H\frac{H_{y0}}{H_{x0}}\cdot\frac{\lambda}{\delta}\,\bar H_y\frac{\partial\bar H_x}{\partial\bar y} =$$

$$1 \qquad 1 \qquad\qquad \frac{1}{\delta}\quad \delta \qquad\qquad \delta \quad \frac{1}{\delta}$$

$$-\frac{p_0}{\rho_0 U^2\bar\rho}\cdot\frac{\partial\bar p}{\partial\bar x} - R_H\frac{H_{y0}^2}{H_{x0}^2}\cdot\frac{\bar H_y}{\bar\rho}\cdot\frac{\partial\bar H_y}{\partial\bar x} + \frac{1}{R_e}\cdot\frac{1}{\bar\rho} \tag{12.57}$$

$$\delta^2 \qquad\qquad\qquad \delta^2$$

$$\left\{\bar\mu\left[2\frac{\partial\bar u}{\partial\bar x} - \tfrac{2}{3}\left(\frac{\partial\bar v}{\partial\bar y}\cdot\frac{\lambda}{\delta}\cdot\frac{V}{U} + \frac{\partial\bar u}{\partial\bar x}\right)\right] + \frac{\lambda}{\delta}\cdot\frac{\partial}{\partial\bar y}\left[\bar\mu\left(\frac{\partial\bar u}{\partial\bar y}\cdot\frac{\lambda}{\delta} + \frac{\partial\bar v}{\partial\bar x}\cdot\frac{V}{U}\right)\right]\right\}$$

$$1 \qquad\qquad \frac{1}{\delta}\quad\delta\quad 1 \qquad \frac{1}{\delta} \qquad\qquad \frac{1}{\delta} \qquad\qquad \delta$$

$$\frac{\partial\bar v}{\partial\bar t}\cdot\frac{V}{U} + \bar u\frac{\partial\bar v}{\partial\bar x}\cdot\frac{V}{U} + \bar v\frac{\partial\bar v}{\partial\bar y}\cdot\frac{\lambda}{\delta}\left(\frac{V}{U}\right)^2 - R_H\frac{H_{v0}}{H_{x0}}\cdot\frac{\bar H_x}{\bar\rho}\cdot\frac{\partial\bar H_y}{\partial\bar x} =$$

$$\delta \qquad\qquad \delta \qquad\qquad \frac{1}{\delta}\quad\delta^2 \qquad\qquad \delta$$

$$-\frac{p_0}{\rho_0 U^2\bar\rho}\cdot\frac{\partial\bar p}{\partial\bar y}\cdot\frac{\lambda}{\delta} - R_H\frac{\bar H_x}{\bar\rho}\cdot\frac{\partial\bar H_x}{\partial\bar y}\cdot\frac{\lambda}{\delta} + \frac{1}{R_e}\cdot\frac{1}{\bar\rho} \tag{12.58}$$

$$\frac{1}{\delta} \qquad\qquad\qquad \frac{1}{\delta}\quad \delta^2$$

$$\left\{\frac{\lambda}{\delta}\cdot\frac{\partial}{\partial\bar y}\left[\bar u\left(2\frac{\partial\bar v}{\partial\bar y}\cdot\frac{V}{U}\cdot\frac{\lambda}{\delta} - \tfrac{2}{3}\left(\frac{\partial\bar u}{\partial\bar x} + \frac{\partial\bar v}{\partial\bar y}\cdot\frac{V}{U}\cdot\frac{\lambda}{\delta}\right)\right)\right] + \frac{\partial}{\partial\bar x}\left[\bar u\left(\frac{\partial\bar u}{\partial\bar y}\cdot\frac{\lambda}{\delta} + \frac{\partial\bar v}{\partial\bar x}\cdot\frac{V}{U}\right)\right]\right\}$$

$$\frac{1}{\delta} \qquad \delta\quad\frac{1}{\delta} \qquad \delta \qquad \delta\quad\frac{1}{\delta} \qquad\qquad \frac{1}{\delta} \qquad \delta$$

where $R_H = \mu_e H_{x0}^2/\rho_0 U^2$ = magnetic pressure number which is of the order of unity and $R_e = U \lambda \rho_0/\mu_0$ = Reynolds number.

The non-dimensional form of the energy equation is

$$\frac{C_{p0}T_0}{U^2}\bar{C}_p\bar{\rho}\left(\frac{\partial\bar{T}}{\partial\bar{t}}+\bar{u}\frac{\partial\bar{T}}{\partial\bar{x}}+\bar{v}\frac{\partial\bar{T}}{\partial\bar{y}}\cdot\frac{V}{U}\cdot\frac{\lambda}{\delta}\right)$$

$$1 \qquad 1 \qquad 1 \qquad \delta \quad \frac{1}{\delta}$$

$$=\frac{p_0}{\rho_0 U^2}\left(\frac{\partial\bar{p}}{\partial\bar{t}}+\bar{n}\frac{\partial\bar{p}}{\partial\bar{x}}+\bar{v}\frac{\partial\bar{p}}{\partial\bar{y}}\cdot\frac{V}{U}\cdot\frac{\lambda}{\delta}\right)+\frac{R_H}{R_\sigma}\bar{\nu}_H\left(\frac{H_{y0}}{H_{x0}}\cdot\frac{\partial\bar{H}_y}{H\bar{x}}-\frac{\lambda}{\delta}\cdot\frac{\partial\bar{H}_x}{\partial\bar{y}}\right)^2$$

$$1 \qquad 1 \qquad 1 \qquad \delta \quad \frac{1}{\delta} \qquad \delta^2 \qquad \delta^2 \qquad \frac{1}{\delta^2} \tag{12.59}$$

$$+\frac{1}{P_r R_e}\cdot\frac{T_0 C_{p0}}{U^2}\left[\frac{\partial}{\partial\bar{x}}\left(\bar{\varkappa}\frac{\partial\bar{T}}{\partial\bar{x}}\right)+\frac{\partial}{\partial\bar{y}}\left(\bar{\varkappa}\frac{\partial\bar{T}}{\partial\bar{y}}\right)\cdot\left(\frac{\lambda}{\delta}\right)^2\right]+\frac{1}{R_e}\left\{\bar{\mu}\left[2\left(\frac{\partial\bar{u}}{\partial\bar{x}}\right)^2+\right.\right.$$

$$1 \quad \delta^2 \qquad 1 \qquad 1 \qquad \frac{1}{\delta^2} \qquad \delta^2 \qquad 1$$

$$\left.\left.+2\left(\frac{\partial\bar{v}}{\partial\bar{y}}\right)^2\cdot\left(\frac{V}{U}\right)^2\cdot\left(\frac{\lambda}{\delta}\right)^2+\left(\frac{\partial\bar{v}}{\partial\bar{x}}\cdot\frac{V}{U}+\frac{\partial\bar{u}}{\partial\bar{y}}\cdot\frac{\lambda}{\delta}\right)^2-\tfrac{2}{3}\left(\frac{\partial\bar{u}}{\partial\bar{x}}+\frac{\partial\bar{v}}{\partial\bar{y}}\cdot\frac{V}{U}\cdot\frac{\lambda}{\delta}\right)^2\right]\right\}$$

$$\delta^2 \qquad \frac{1}{\delta^2} \qquad \delta^2 \qquad \frac{1}{\delta^2} \qquad 1 \qquad \delta \quad \frac{1}{\delta}$$

where the Prandtl number $P_r = C_{p0}\,\mu_0/\varkappa_0$ is of the order of unity. The quantities $C_{p0}\,T_0/U^2$ and $p_0/P_0\,U^2$ are of the order of unity.

If we neglect the terms of the order of δ and smaller in equations (12.53) to (12.59), we have the following boundary layer equations for the two-dimensional flow of a plasma with large magnetic Reynolds number over a flate plate:

$$\frac{\partial H_x}{\partial t}+u\frac{\partial H_x}{\partial x}+v\frac{\partial H_x}{\partial y}+H_x\frac{\partial v}{\partial y}-H_y\frac{\partial u}{\partial y}=\frac{\partial}{\partial y}\left(\nu_H\frac{\partial H_x}{\partial y}\right) \tag{12.60}$$

$$\frac{\partial H_y}{\partial t}+u\frac{\partial H_y}{\partial x}+v\frac{\partial H_y}{\partial y}-H_x\frac{\partial v}{\partial x}+H_y\frac{\partial u}{\partial x}=-\frac{\partial}{\partial x}\left(\nu_H\frac{\partial H_x}{\partial y}\right) \tag{12.61}$$

$$\frac{\partial u}{\partial t}+u\frac{\partial u}{\partial x}+v\frac{\partial u}{\partial y}-\frac{\mu_e}{\rho}H_y\frac{\partial H_x}{\partial y}=-\frac{1}{\rho}\cdot\frac{\partial p}{\partial x}+\frac{1}{\rho}\cdot\frac{\partial}{\partial y}\left(\mu\frac{\partial u}{\partial y}\right) \tag{12.62}$$

$$\frac{\partial}{\partial y}\left(p+\frac{\mu_e H_x^2}{2}\right)=0 \tag{12.63}$$

$$\rho\left(\frac{dh}{\partial t}+u\frac{\partial h}{\partial x}+v\frac{\partial h}{\partial x}\right)=\frac{\partial p}{\partial t}+u\frac{\partial p}{\partial x}+v\frac{\partial p}{\partial y}+\frac{\partial}{\partial y}\left(\varkappa\frac{\partial T}{\partial y}\right)+\mu\left(\frac{\partial u}{\partial y}\right)^2+\nu_H\left(\frac{\partial H_x}{\partial y}\right)^2 \tag{12.64}$$

$$\frac{\partial\rho}{\partial t}+\frac{\partial\rho u}{\partial x}+\frac{\partial\rho v}{\partial y}=0 \tag{12.65}$$

$$p=\rho R_p T \tag{12.66}$$

where $h = C_p\,T$.

It is easy to show that equation (12.61) may be replaced by equation (12.50). Integration of equation (12.63) gives

$$p + \frac{\mu_e H_x^2}{2} = p_0(x) \tag{12.67}$$

where $p_0(x)$ is the pressure just outside of the boundary layer and is assumed to be known in the analysis of boundary layer flow. For ordinary boundary layer flow, the pressure in the boundary layer is equal to the given function $p_0(x)$. In the present case, the sum of the gasdynamic pressure p and the magnetic pressure $\mu_e H_x^2/2$ is equal to the given pressure function $p_0(x)$.

It should be noticed that in general there are three different boundary layers in magnetogasdynamics, i. e., the boundary layer of velocity profile whose thickness δ_v is proportional to $\sqrt{R_e}$, the thermal boundary layer of temperature profile whose thickness δ_T is proportional to $\sqrt{P_r R_e}$ and the boundary layer of magnetic field profile whose thickness δ_H is proportional to $\sqrt{R_\sigma}$. Equation (12.46) represents the case $\delta_H \gg \lambda \gg \delta_v \cong \delta_T$. Equations (12.60) to (12.66) represent the case $\delta_v \cong \delta_T \cong \delta_H \geqslant \lambda$. For a plasma $P_r = 0\,(1)$, hence $\delta_v \cong \delta_T$. However there is another interesting case in which $1 \ll R_\sigma \ll R_e$ so that $\lambda \gg \gg \delta_H \gg \delta_v$. In this case, in the boundary layer of velocity profile, we may use equation (12.46). In the boundary layer of magnetic field, we may assume that the plasma is inviscid and non-heat-conducting. It is possible to show that under certain condition, this boundary layer region of magnetic field is a cavity in which the flow velocity is zero. We may have such an interesting case in which a free streamline will cover a magnetized body to form a cavity over this body where the incident flow is assumed to be absent of magnetic field (*4*).

References

1. Bush, W. B.: Magnetohydrodynamic-Hypersonic Flow Past a Blunt Body. Jour. Aero-Space Sci, vol. 25, No. 11, Nov. 1958, pp. 685–690.

2. Friedrichs, K. O.: Nonlinear Wave Motion in Magneto-Hydrodynamics. Los Almos Laboratory Report, 1954.

3. Friedrichs, K. O., and H. Kranzer: Nonlinear Wave Motion. Notes on Magnetohydrodynamics VIII, Institute of Math. Sci., New York Univ., 1958.

4. Kulikovskii, A. G.: On the Flow of a Conducting Fluid Around a Magnetized Body. Doklady, AN USSR, vol. 117, No. 2, 1957, pp. 199–202.

5. McGune, J. E., and E. L. Resler Jr.: Compressibility Effects in Magneto-aerodynamic Flows past thin Bodies. Jour. Aero-Space Sci., vol. 27, No. 7, July 1960, pp. 493–503.

6. Pai, S. I.: Viscous flow theory, I—Laminar flow. D. Van Nostrand Co., Inc., Princeton, N. J., 1956.

7. Pai, S. I.: Linearized Theory of Airfoils in Fluids of Low Electric Conductivity. Tech. Inf. Series R 60 SD 311, Space Sciences Lab. General Electric Co., Dec. 1959.

8. Sears, W. R., and E. L. Resler Jr.: Sub- and Super-Alfvenic Flows Past Bodies. Preprint, Second Intern. Congress of the Aero. Sci., Zurich, Switzerland, Sept. 1960.

Chapter XIII

Some Problems of Plasma Dynamics

1. Introduction. In chapters VI to XII, we considered the plasma as a single fluid of definite composition. Actually the composition, particularly the degree of ionization of the plasma, changes with temperature. If the temperature of the plasma varies through a large range in the flow field, the change of the composition of the plasma may be important. In this chapter we shall study some simple problems of plasma dynamics where the composition of the plasma may not be considered as unchanged.

The most important factor for the change of the composition of a plasma is its degree of ionization. In § 2, we are going to discuss briefly the ionization process with special reference to plasma dynamics. In studying the flow problems, usually we are not interested in the detailed process of ionization but in the degree of ionization of a plasma for a given flow condition. In other words, we would like to know the degree of ionization of plasma at a given temperature and pressure. This is given by the well-known Saha relation which will be discussed in § 3.

In § 4, we shall discuss the methods of analysis of the flow of plasma with variable degree of ionization while in § 5, we shall study the plasma oscillation where the plasma is not considered as a single fluid as we did in Chapter X.

One of the most important factors in plasma dynamics when the plasma is considered as a continuum is the concept of electrical conductivity. In Chapters VI to XII, we considered the electrical conductivity as a scalar quantity. As we have shown in chapter III, § 4, the electrical conductivity is really a tensor quantity. In § 6, we shall consider further the electrical conductivity and its influence on flow problems.

Another important factor which we have omitted in our study of the flow problems in chapters VI to XII is the radiation phenomenon. At very high temperature, the radiation phenomenon may be very important and it should be taken into account in the study of the flow problem. In § 7, we shall briefly treat a simple flow problem with radiation term.

If the plasma is rarefied, the theory of continuum may not hold true. In such a case, we should use the kinetic theory to treat the flow of plasma. We shall discuss briefly the kinetic theory of plasma in Chapter XIV. For an extremely rarefied plasma, we may treat the particles of plasma as discrete point masses in space under the influence of electromagnetic fields by neglecting completely the interaction between particles. Some interesting phenomena may be obtained in this manner. We shall treat such problems in § 8.

2. Ionization. For simplicity, we shall assume that the plasma consists entirely of atoms. This is a good approximation at very high temperatures.

Ionization is a process in which an electron is removed from an atom, leaving the atom with a net positive charge. Ionization is closely related to the structure of the atom. The following two terms are used in the analysis of the ionization process:

(i) First ionization potential V_i of the atom. This is the energy required to remove completely an outer electron from its normal state in a neutral atom to a distance beyond the sphere of influence of the nucleus.

(ii) First resonance potential V_R of the atom. This is the energy required to raise the outer electron from its normal level to the next higher energy level.

Both of the above energies are usually expressed in electron volts. The energy is simply $V_i e$ or $V_R e$ where e is the absolute electric charge.

The energy necessary for excitation or ionization may be given to an atom, by an electron collision, a positive ion collision, the absorption of a quantum of radiant energy, or collision of neutral atoms at high temperature. In a plasma the temperature is high and the ionization is essentially due to thermal ionization. Thermal ionization is a combination of ionization processes due to molecular collisions, radiation and electron collisions occurring in the plasma at high temperature. In reference (*3*), a detailed analysis of these three processes involved in thermal ionization has been given. We shall summarize those results as follows (*18*).

(1) *Ionization by thermal collision.* At a very high temperature, the kinetic energy of the atoms is very high. As a result, many atoms in the plasma will have sufficient energy to ionize the atoms they strike in a collision. If we assume that the velocity distribution is Maxwellian, the rate of increase of the number density of the ions is

$$\frac{d\,\nu_i}{d\,t} = \frac{4}{\sqrt{2\,m}} \cdot \frac{f_i\,\nu_0}{\lambda} \cdot \frac{(e\,V_i)^{\frac{5}{2}}}{(k\,T)^2} \exp\left(-\frac{2\,e\,V_i}{k\,T}\right) \tag{13.1}$$

where ν_i is the number density of the ions, ν_0 is the total number density of the original neutral atoms in the plasma, m is the mass of the atom, λ is the mean free path which may be written as $\lambda = 1/S_i\,\nu_0\sqrt{2}$, S_i is the cross section for ionization, k is the Boltzmann constant, T is the absolute temperature and f_i is the ratio of the number of ionizing collisions to the number of collisions and sufficient energy. The value of f_i is 10^{-2} to 10^{-4}.

(2) *Photonionization.* At very high temperatures the radiation energy is large. If the energy of a photon is greater than $e\,V_i$ of the atom, ionization takes places when the atom absorbs the radiant energy of this photon. This process is known as photonionization. If we assume black body radiation in the plasma, the rate of increase of the number of ions due to photoionization is

$$\frac{d\,\nu_i}{d\,t} = \frac{2\,\pi\,k^3}{c^2} \cdot \frac{T^3}{h^2}\,\mu_p \exp\left[\left(-\frac{e\,V_i}{k\,T}\right)^2 + \frac{2\,e\,V_i}{k\,T} + 2\right] \tag{13.2}$$

where μ_p is the coefficient for absorption of photons, c is the velocity of light and h is Planck's constant.

(3) *Ionization by electron collision.* The previous processes produce electrons which in thermal equilibrium must have the same kinetic energy as atoms or

ions in the plasma. Since the mass of the electron is much smaller than that of the atom or ions, its velocity will be very high. These electrons in collision with neutral particles cause ionization. The rate of increase of the number density of ions by electron collision is

$$\frac{d\,\nu_i}{d\,t} = 600\, a_1 \frac{m\,p}{e\sqrt{\pi}}\,\nu_e \left(\frac{2\,k\,T}{m}\right)^{\frac{3}{2}} \exp\left(-\frac{e\,V_i}{k\,T}\right)\cdot\left(1+\frac{e\,V_i}{2\,k\,T}\right) \tag{13.3}$$

where p is the pressure in mm. Hg., ν_e is the number density of the electron and a_1 is a factor depending on the gas. For air $a_1 = 0.26$.

Equations (13.1) to (13.3) give us the rate of ionization by three different process. In actual cases, ionization of an atom may be completed by more than one process (*18*). An atom may first be excited by one process and then ionized by another process. This is known as cumulative ionization. For instance an atom may be excited by absorption of a photon and this excited atom may be ionized by collision with other atoms.

In a plasma, the net increase of ions is the difference of the rate of increase of ions by ionization processes and the rate of recombination of ions and electrons. The rate of recombination may be written as follows:

$$\frac{d\,\nu_t}{d\,t} = -\,\beta_i\,\nu_i\,\nu_e \tag{13.4}$$

where β_i is known as the recombination coefficient.

If there are walls around the plasma, the influence of the wall on ionization should also be considered. Hence for actual flow conditions, ionization presents a very complicated problem (*8*).

3. Saha relation (*14*). In plasma dynamics, we are not interested in the detailed processes of ionization but in the degree of ionization for a given flow condition. In other words, we would like to know the degree of ionization of the plasma at a given temperature and pressure. SAHA was the first one to make a successful analysis of the degree of ionization of a plasma in an equilibrium state (*14*). He considered the dissociation of neutral particles into positive ions and electrons and applied the reaction equation of thermodynamics, sometimes known as the law of mass action (*4*), to the ionization process, i. e.,

one mol of atoms + ionization energy for one mol of neutral atoms =
= one mol of ions + one mol of electrons. (13.5)

The law of mass action (*4*) or equation (13.5) gives the following relation:

$$\frac{p_i\,p_e}{p_a} = C\,T^{\frac{5}{2}} \exp\left(-\frac{e\,V_i}{k\,T}\right) \tag{13.6}$$

where p_i is the partial pressure of ions, p_e is the partial pressure of electrons and p_a is the partial pressure of neutral atoms. All the three species are considered as ideal gases. The constant C must be obtained from statistical mechanics and is given by (*12*)

$$C = \frac{(2\,\pi\,\mu_0)^{\frac{3}{2}}\,k^{\frac{5}{2}}}{h^3}\cdot\frac{Z_i\,Z_e}{Z_a} \tag{13.7}$$

where $\mu_0 = m_i\, m_e/m_a$ = the reduced mass $\simeq m_e$. Z is the internal partition function defined by

$$Z = \sum_n g_n \exp\left(-\frac{E_n}{k\,T}\right) \tag{13.8}$$

where E_n and g_n are the energy and the statistical weight, respectively, of the nth state of the system. E_0 is taken as zero. The partition function for the electrons is 2 and that for the atoms is 1. The partition function for ions may be taken as $4 + 2\exp(-E_1/k\,T)$ as a first approximation.

In terms of degree of ionization α and total pressure of the plasma p, if there is only one kind of neutral atoms, equation (13.6) becomes

$$\frac{\alpha^2}{1-\alpha^2} = C\,\frac{T^{\frac{5}{2}}}{p}\exp\left(-\frac{e\,V_i}{k\,T}\right) \tag{13.9}$$

where α = degree of ionization $= \nu_i/\nu_0$.

Sometimes it is convenient to write equation (13.9) in the following form:

$$\log_{10}\left(\frac{\alpha^2}{1-\alpha^2}\,p\right) = 2.5\log_{10} T - \frac{5040\,V_i}{T} + C_1 \tag{13.10}$$

where p is the total pressure of the plasma in mm. Hg., T is the absolute temperature in degrees Kelvin, V_i is in volts and $C_1 = \log_{10}$ C. The constant C_1 is called the chemical constant which depends on the excited states of the neutral particles as given in equation (13.7). For a first approximation as suggested by ENGEL and STEENBECK (*3*), $C_1 = -3.62$ for unexcited neutral atoms and singly charged ions. For more accurate calculations, equation (13.7) should be used. For instance, SCHEIBE (*15*) gave the following equation for the degree of ionization of Argon:

$$\alpha = \left[\frac{p}{T^{\frac{5}{2}}}\cdot\frac{10^4}{2+\exp\left(-\frac{2060}{T}\right)}\exp\left(\frac{182000}{T}\right) + 1\right]^{-\frac{1}{2}} \tag{13.11}$$

where p is in cm. Hg., $E_1/k = 2060°$, $e\,V_i/k = 182.000°$ for Argon.

For a mixture of different kinds of neutral atoms, each with a different ionization potential, we can not use equation (13.10) to determine the degree of ionization of the rth species by simple using the partial pressure ot the rth species in the plasma. This is due to the fact that in equilibrium the electrons produced from one species depend also on the conditions of all the other species in the plasma. The actual degree of ionization depends on the equilibrium condition of the system as a whole. Let α_r be the degree of ionization of the rth species, i. e.,

$$\alpha_r = \frac{\text{no. density of ions of rth species}}{\text{no. dencity of original rth neutral atoms}} = \frac{\nu_i(r)}{\nu_0(r)} \tag{13.12}$$

The mean degree of ionization of the plasma as a whole is then

$$\alpha_m = \frac{\sum_r \nu_i(r)}{\sum_r \nu_0(r)} \tag{13.13}$$

The Saha equation for α_r becomes

$$\log_{10}\left(\frac{\alpha_r}{1-\alpha_r}\cdot\frac{\alpha_m}{1+\alpha_m}p\right) = 2.5\log_{10} T - \frac{5040\, V_i(r)}{T} + C_1 \qquad (13.14)$$

Equation (13.14) should be used instead of equation (13.10). If there are q different kinds of neutral atoms, we have q equations of the type equation (13.14). We have to solve $q+1$ equations, i. e., equation (13.13) and q equations of the type of equation (13.14) for the values of α_m and α_r, $r = 1, \ldots q$. In general, the atoms having the lowest ionization potential in the mixture will be most ionized and the atoms having the highest ionization potential will be least ionized.

4. Flow of plasma with variable degree of ionization. There are two different methods of analysis of flow problems of plasma with a variable degree of ionization:

(1) The accurate method is to consider the plasma as a mixture of neutral particles, ions and electrons. We are going to find the partial variables of all species in the plasma by equations discussed in chapter II, § 9. The production terms σ_r may be calculated by the results of the ionization process discussed in § 2. We shall use this method to discuss some simple plasma oscillations in the next section.

(2) An approximate method is to use the theory of a single fluid but with the degree of ionization α as an additional variable. The additional relation is the Saha relation of equation (13.10). The influences of the degree of ionization α are as follows:

(i) The equation of state should be rewritten in the following form:

$$p = (1+\alpha)\, R_g\, \rho\, T \qquad (13.15)$$

where R_g is the gas constant of the unionized gas.

(ii) The internal energy of the gas depends on the degree of ionization. For instance, the internal energy of an ionized gas of a single charge is

$$U = \frac{R_g T}{\gamma - 1}(1+\alpha) + \alpha\, R_g\left(\frac{e\, V_i}{k}\right) \qquad (13.16)$$

(iii) The transport coefficients (*1*) such as the coefficient of viscosity, the coefficient of thermal conductivity and the electrical conductivity are functions of α. The dependence of the electrical conductivity of the degree of ionization is of utmost importance and will be discussed in § 6.

Let us calculate the Rankine-Hugoniot relation of a strong shock wave of speed U_s traveling into a medium at rest of pressure p_1, temperature T_1 and density ρ_1. We are going to calculate the degree of ionization, the pressure p_2, the temperature T_2, the density ρ_2 and the speed of the flow u_2 behind the shock wave. The conservation laws across the shock are:

$$\rho_1 U_s = \rho_2 (U_s - u_2) \qquad (13.17)$$

$$p_1 + \rho_1 U_s^2 = p_2 + \rho_2 (U_s - u_2)^2 \qquad (13.18)$$

$$H_1 + \tfrac{1}{2} U_S^2 = H_2 + \tfrac{1}{2}(U_S - u_2)^2 \qquad (13.19)$$

where $H = U + p/\rho$ and subscript 1 refers to the values in front of the shock and subscript 2 refers to those behind the shock. The internal energy U is given by equation (13.16). We are going to obtain a method of solution for the five unknown p_2, T_2, ρ_2, α and u_2 in terms of the known quantities p_1, T_1, ρ and U_s. Here we assume that $\alpha_1 = 0$. The method of solution is exactly the same if α_1 is different from zero. The five relations for these unknows are equations (13.17) to (13.19) and (13.15) and (13.10). Because of the complicated form of equation (13.10), it is not possible to solve these unknowns analytically. An iteration procedure should be used. We may first assume an arbitrary value of α for a given U_s and solve for p_2, T_2, ρ_2 and u_2. With the known T_2 and p_2 we may calculate α from equation (13.10). If the value of α given by equation (13.10) is the same as our assumed value, our final solution is obtained. If the assumed and the computed values of α are not the same, we have to repeat the procedure until the two values of α check.

In general the effect of ionization is to lower the temperature and to raise the density and the velocity behind the shock compared to the case without ionization for the same shock strength and the initial conditions. The pressure behind the shock is also increased slightly by the effect of ionization. The degree of ionization increases rapidly with increasing temperature. For a shock speed of $M_1 = U_s/a_1 = 20$, where $a_1{}^2 = \gamma R T_1$, Scheibe (*15*) found that the degree of ionization of Argon is about 30%.

5. Plasma oscillation. For simplicity we assume that the plasma consists of only ions and electrons. The plasma is at rest with pressure $2\,p_0$, temperature T_0, number density $2\,\nu_0$. There are no external applied electromagnetic fields. The plasma may be considered as invisid and non-heat conducting fluid. The plasma is perturbed by a small disturbance so that in the resultant disturbed motion, we have the following partial variables:

$$\begin{gathered} u_1{}^i = u_1{}^i\,(x, t), \quad u_2{}^i = u_2{}^i\,(x, t), \quad p_1 = p_0 + p'_1\,(x, t), \quad p_2 = p_0 + p'_2\,(x, t) \\ T_1 = T_0 + T'_1\,(x, t), \quad T_2 = T_0 + T_2'\,(x, t), \quad \nu_1 = \nu_0 + \nu'_1\,(x, t), \qquad (13.20) \\ \nu_2 = \nu_0 + \nu_2'\,(x, t), \quad E^i = E^i\,(x, t), \quad H^i = h^i\,(x, t) \end{gathered}$$

where the perturbed quantities $u_1{}^i$, $u_2{}^i$, T'_1, T_2', p'_1, p_2', ρ'_1, ρ_2', E^i and h^i are assumed to be small and the second and higher order terms of these quantities are negligible. We assume that these perturbed quantities are function of x and t only. Thus we will discuss only the wave propagation in the direction of x-axis. Subscript 1 and 2 refers to values for ions and electrons respectively.

Substituting equation (13.20) into the fundamental equations of plasma dynamics discussed in chapter II, § 9 and neglecting the higher order terms, we have following 18 linear equations for the perturbed quantities:

$$\frac{p_1'}{p_0} = \frac{\nu_1'}{\nu_0} + \frac{T_1'}{T_0} \tag{13.21}$$

$$\frac{p_2'}{p_0} = \frac{\nu_2'}{\nu_0} + \frac{T_2'}{T_0} \tag{13.22}$$

$$\frac{\partial \varepsilon E^1}{\partial t} + e \nu_0 (u_1{}^1 - u_2{}^1) = 0 \tag{13.23}$$

$$\frac{\partial \varepsilon E^2}{\partial t} + e \nu_0 (u_1{}^2 - u_2{}^2) = -\frac{\partial h^3}{\partial x} \tag{13.24}$$

$$\frac{\partial \varepsilon E^3}{\partial t} + e \nu_0 (u_1{}^3 - u_2{}^3) = \frac{\partial h^2}{\partial x} \tag{13.25}$$

$$\frac{\partial \mu_e h^1}{\partial t} = 0 \tag{13.26}$$

$$\frac{\partial \mu_e h^2}{\partial t} = \frac{\partial E^3}{\partial x} \tag{13.27}$$

$$\frac{\partial \mu_e h^3}{\partial t} = -\frac{\partial E^2}{\partial x} \tag{13.28}$$

$$\frac{\partial \nu_1'}{\partial t} + \nu_0 \frac{\partial u_1{}^1}{\partial x} = 0 \tag{13.29}$$

$$\frac{\partial \nu_2'}{\partial t} + \nu_0 \frac{\partial u_2{}^1}{\partial x} = 0 \tag{13.30}$$

$$m_1 \nu_0 \frac{\partial u_1{}^1}{\partial t} = -\frac{\partial p_1'}{\partial x} + e \nu_0 E^1 + \alpha_{12} (u_1{}^1 - u_2{}^1) \tag{13.31}$$

$$m_1 \nu_0 \frac{\partial u_1{}^2}{\partial t} = e \nu_0 E^2 + \alpha_{12} (u_1{}^2 - u_2{}^2) \tag{13.32}$$

$$m_1 \nu_0 \frac{\partial u_1{}^3}{\partial t} = e \nu_0 E^3 + \alpha_{12} (u_1{}^3 - u_2{}^3) \tag{13.33}$$

$$m_2 \nu_0 \frac{\partial u_2{}^1}{\partial t} = -\frac{\partial p_2'}{\partial x} - e \nu_0 E^1 + \alpha_{12} (u_2{}^1 - u_1{}^1) \tag{13.34}$$

$$m_2 \nu_0 \frac{\partial u_2{}^2}{\partial t} = -e \nu_0 E^2 + \alpha_{12} (u_2{}^2 - u_1{}^2) \tag{13.35}$$

$$m_2 \nu_0 \frac{\partial u_2{}^3}{\partial t} = -e \nu_0 E^3 + \alpha_{12} (u_2{}^3 - u_1{}^2) \tag{13.36}$$

$$m_1 \nu_0 C_{p1} \frac{\partial T_1'}{\partial t} = \frac{\partial p_1'}{\partial t} \tag{13.37}$$

$$m_2 \nu_0 C_{p2} \frac{\partial T_2'}{\partial t} = \frac{\partial p_2'}{\partial t} \tag{13.38}$$

where the superscripts 1, 2 and 3 refer to the components along x-, y- and z-axis respectively. The interaction force between ions and electrons is characterized by the friction coefficient α_{12} [cf. equation (3.19)]. The absolute electric charge is e. It is interesting to notice that the variables may be divided into four groups:

(i) The x-component of magnetic field h^1 given by equation (13.26) is independent of all the other variables. We may take it as zero.

(ii) The second group consists of u_1^2, u_2^2, E^2, h^3 which are related by equations (13.24), (13.28), (13.32) and (13.35).

(iii) The third group consists of u_1^3, u_2^3, E^3 and h^2 which are related by equations (13.25), (13.27), (13.33) and (13.36).

(iv) The last group consists of the rest of the variables.

The equations for the second and third group are identical in form. These two groups characterize a transverse wave. The last group is a longitudinal wave. Just as in the results of the wave motion of a single fluid discussed in chapter X, we see that the transverse wave is independent of the compressibility effect of the plasma while the longitudinal wave depends on the compressibility of the plasma.

If we are looking for periodic solutions in which all the perturbed quantities are proportional to

$$e^{i\omega t - i\lambda x} \tag{13.39}$$

where ω is a given real quantities and $i = \sqrt{-1}$. The determinantal equation of each group gives the corresponding dispersion relation for the corresponding group, i. e., the relation between λ and ω and all other physical quantities such as ν_0, α_{12} etc.

For the transverse wave, we have the following relation (*8*):

$$c^2 \lambda^2 = \omega^2 - \omega_e^2 \left(\frac{1 - i\,\alpha_{12}^*}{1 + i\,\alpha_{12}^*}\right) \tag{13.40}$$

where

$$\omega_e = e \sqrt{\frac{\nu_0}{\varepsilon\, m_2}} = \text{electron plasma frequency} \tag{13.41}$$

$$\alpha_{12}^* = \frac{\alpha_{12}}{m_2\, \omega\, \nu_0} \tag{13.42}$$

The friction coefficient α_{12} is closely related to the electric conductivity σ, i. e.,

$$\sigma = \frac{e^2\, \nu_0^2}{\alpha_{12}} \tag{13.43}$$

For infinite electric conductivity, $\alpha_{12} = 0$. In this case, equation (13.40) becomes

$$\lambda^2 = \frac{\omega^2}{c^2}\left(1 - \frac{\omega_e^2}{\omega^2}\right) \tag{13.44}$$

In an ideal plasma, we have undamped transverse wave when $\omega > \omega_e$ and damped exponential transverse wave when $\omega < \omega_e$. The characteristic frequency ω_e is the well-known plasma frequency of electrons.

For the longitudinal waves, we have the following relations (*8*):

$$\lambda^4 - \frac{\lambda^2\,\omega^2}{a_1^2}\left\{\left(1 + 2\,i\,\alpha_{12}^*\,\frac{m_2}{m_1}\right) - 2\,\frac{m_2\,\omega_e^2}{m_1\,\omega^2}\right\} + \frac{\omega^4}{a_1^2\,a_2^2}\left\{(1 + i\,\alpha_{12}^*) - \frac{\omega_e^2}{\omega^2}\right\} = 0 \tag{13.45}$$

where

$$a_1 = \sqrt{\frac{\gamma p_0}{m_1 \nu_0}} = \text{ion sound speed} \tag{13.46a}$$

$$a_2 = \sqrt{\frac{\gamma p_0}{m_2 \nu_0}} = \text{electron sound speed} \tag{13.46b}$$

$$a_p = \sqrt{2}\, a_1 = \text{plasma sound speed} \tag{13.46c}$$

There are two roots of λ^2 from equation (13.45) which represent two different modes of longitudinal waves. The first mode represents an ion wave for which λ^2 in the case of ideal plasma is always positive. At low applied frequency ω, the speed of propagation of this mode is the sound speed of the plasma as a whole, i. e., a_p. This speed of wave propagation decreases as ω increases. At very high frequency, it is equal to the ion sound speed a_1. The second mode represents an electron wave for which λ^2 is negative for $\omega < \omega_e$ and positive for $\omega > \omega_e$ in the case of ideal plasma. Hence at low frequency, the second mode is a damped exponential wave while at high frequency $\omega > \omega_e$, it is an undamped wave. At very high frequency, the speed of propagation of the second mode is the sound speed of the electrons.

The effect of finite electric conductivity, i. e., $\alpha_{12} \neq 0$ is to introduce damping in the undamped sound speed.

6. Electrical conductivity. One of the most important quantities in plasma dynamics is the electrical conductivity σ, which may be defined by equation (3.24), i. e., the Joule heat is given by the relation:

$$\text{Joule heat} = i^2/\sigma \tag{3.24}$$

The inverse of electrical conductivity is called the electrical resistivity R, i. e., $R = 1/\sigma$. If we define the electrical conductivity σ by equation (3.22), i. e.,

$$\vec{i} = \sigma\, \vec{E}_u \tag{3.22}$$

We have shown that the electrical conductivity is a tensor quantity when there is a magnetic field (see chapter III, § 4).

In our simple analysis of chapter III, § 4, we find that the electrical conductivity σ is given by the following formula

$$\sigma = -\frac{\rho_{e1}^2}{\alpha_{12}} \tag{13.43}$$

The friction coefficient α_{12} should be calculated from the kinetic theory of plasma (*1*). If we assume that the interaction force is simply due to the momentum gained by electrons in collision with ions, the simple kinetic theory gives

$$F_{ee0} = m_e \nu_e f (u_i - u_e) = \alpha_{12} (u_e - u_i) \tag{13.47}$$

Hence

$$\alpha_{12} = -m_e \nu_e f \tag{13.48}$$

where f is the collision frequency of each electron. We have then

$$\sigma = \frac{e^2 \nu_e}{f\, m_e} \tag{13.43a}$$

For a more accurate analysis, we must use the generalized Boltzmann equation [cf. chapter XIV (14.8)] to determine the electrical conductivity. There are two types of collision in plasma, i. e., the close encounters between the electrons and the neutral atoms and the distant encounters between the ions. The theories of electrical conductivity in plasma have generally been developed when only one of these two collision processes exists (*6*).

For a slightly ionized gas, the close encounters dictates the electron mobility. Chapman and Cowling (*2*) gave the following formula for the electrical conductivity for this case with the assumption of rigid spherical molecules:

$$\sigma = 0.532 \frac{\alpha\, e^2}{(m_e\, k\, T)^{1/2}} \cdot \frac{1}{Q} = \sigma_c \tag{13.49}$$

where α is the degree of ionization and Q is the electron-atom-collision cross section.

For fully ionized plasma, the distant encounters between the ions dominate the situation. Spitzer (*14*) and Härm gave the following formula for the electrical conductivity for this case:

$$\sigma = \frac{0.591\, (k\,T)^{\frac{3}{2}}}{m_e^{\frac{1}{2}}\, e^2\, ln \left(\frac{\lambda_d}{b_0}\right)} = \sigma_d \tag{13.50}$$

where λ_d = Debye shielding distance $= \sqrt{(k\,T)/(8\,\pi\,\nu_e\, e^2)}$ and b_0 = impact parameter $= e^2/3\,k\,T$.

For the intermediate case, no theoretical result has been worked out. Kantrowitz (*6*) and his associates suggested the following approximation:

$$\frac{1}{\sigma} = \frac{1}{\sigma_c} + \frac{1}{\sigma_d} \tag{13.51}$$

Under the influence of magnetic field, the electrical conductivity is a tensor quantity as defined by equation (3.23), i. e.,

$$\vec{i} = \sigma\, \vec{E}_{u^{\parallel}} + \frac{\sigma}{1 + \sigma^2 B^2 \xi^2} \vec{E}_{u^{\perp}} - \frac{\xi\, \sigma^2 B}{1 + \sigma^2 B^2 \xi^2} \vec{E}_{u^{\perp}} \times \vec{B}_1 \tag{13.52}$$

where $\vec{B}_1$ is the unit vector in the direction of $\vec{B}$.

The electrical conductivity for the component of $\vec{E}_u$ in the direction of $\vec{B}$ is the same as that in the case of no magnetic field. The electrical conductivity for the component of $\vec{E}_u$ perpendicular to $\vec{B}$ is

$$\sigma^{\perp} = \frac{\sigma}{1 + \sigma^2 B^2 \xi^2} = \frac{\sigma}{1 + \left(\frac{\omega_c}{f}\right)^2} \tag{13.53}$$

where $\omega_c = e\,B/m_e$ = Larmor frequency = cyclotron frequency of electron. In equation (13.53), the simple relation (13.48) is used. In general $\sigma^{\perp}$ is less than σ. Only if $(\omega_c/f) \ll 1$, $\sigma^{\perp}$ is approximately equal to σ.

The last term of (13.52) is the Hall current and the corresponding electrical conductivity is

$$\sigma_h = -\frac{\xi \sigma^2 B}{1 + \sigma^2 B^2 \xi^2} = \frac{\omega_c}{f} \sigma^{\perp} \tag{13.54}$$

For dense plasma and for moderate magnetic field, $(\omega_c/f) \ll 1$, the Hall current may be neglected and the electrical conductivity may be assumed to be a scalar quantity. Otherwise the tensor property of the electrical conductivity should be taken into account.

Equation (13.52) is based on a fully ionized plasma. For a partially ionized plasma, we should use equation (3.26) to determine the tensor property of the electrical conductivity (*5*). At low densities, it can be shown that electrical conductivity σ for the component of $\vec{E}_u{}^{\perp}$ not parallel to $\vec{B}$ is small than that in the direction of $\vec{B}$, i. e., $\sigma^{\perp} < \sigma$.

7. Flow with radiation. At very high temperature which would be usually the case of plasma dynamics, the radiation phenomenon would be important. We have to use the quantum mechanics to study the interaction of a plasma with electromagnetic radiation which is beyond the scope of this book. We shall briefly state the main results of the interaction between matter and radiation in the case of fully ionized plasma as follows (*17*):

There are two main types of interaction between matter and radiation:

(i) A light quantum can be scattered by free electrons. The scattered radiation tends to be polarized.

(ii) A quantum can be emitted or absorbed by electrons in the presence of heavy positive ions. The radiation emitted by an electron in a free-free transition, i. e., the electron remains free in the process, is called Bremsstrahlung. Bremsstrahlung may be a very important source of heat loss at very high temperature. The total amount of energy radiation in Bremsstrahlung is given approximately by the following formula:

$$E_{Rb} = 1.42 \times 10^{-2} Z^3 \nu_1{}^2 T^{\frac{1}{2}} \text{ ergs/cm}^3 \text{ sec} \tag{13.55}$$

where Z is the number of signed elementary charges of ions. For singly charged ions, $Z = 1$.

When an electron emits radiation, it may be captured by the ion. The rates of electron disappearance by radiative recombination is just the reverse process of photonionization.

Since the detailed analysis of radiation effect is very complicated, we may take the radiation effect into consideration of the flow problem approximately by using the radiative equilibrium relations in our fundamental equations of plasma dynamics (*7*) as shown in chapter II. In other words, we include the radiation pressure p_R, radiation energy E_R and radiation flux $Q_R{}^j$ in the fundamental equations of plasma dynamics such as equations (4.3) and (4.4). In this way we may have the first order effect of radiation on the flow field. We are going to consider the following simple flow problem:

We now consider the shock wave structure problem of ordinary gasdynamics but including radiation terms (*10*). The flow is parallel to the x-axis and has the component of velocity u only. For simplicity, we assume that there is no externally applied electromagnetic field and that the forces and energy due to the induced electromagnetic field by motion of the charged particles in the plasma are negligible. Finally we assume that all quantities are functions of x only. We consider that the plasma is a viscous and heat-conducting fluid. Under the above assumptions the fundamental equations of our problem are given below: [cf. equations (4.1) to (4.4)]

$$p = \rho R_p T \tag{13.56}$$

$$\frac{d \rho u}{d x} = 0 \tag{13.57}$$

$$\rho u \frac{d u}{d x} = -\frac{d}{d x}(p + \tfrac{1}{3} a_R T^4) + \tfrac{4}{3} \cdot \frac{d}{d x}\left(\mu \frac{d u}{d x}\right) \tag{13.58}$$

$$\begin{aligned} \rho u \frac{d}{d x}\left(C_v T + \frac{a_R T^4}{\rho}\right) &= \\ = \frac{d}{d y}\left(\varkappa \frac{d T}{d x}\right) - \left(p + \frac{a_R}{3} T^4\right)\frac{d u}{d x} &+ \tfrac{4}{3}\mu\left(\frac{d u}{d x}\right)^2 + \frac{d}{d x}\left(\frac{4 a_R c T^3}{3 \rho \varkappa_R} \cdot \frac{d T}{d x}\right) \end{aligned} \tag{13.59}$$

Integration of equations (13.57) to (13.59) gives respectively

$$\rho u = \rho_0 u_0 = m \tag{13.60}$$

$$m u - m C = -\left(p + \frac{a_R}{3} T^4\right) + \tfrac{4}{3}\mu \frac{d u}{d x} \tag{13.61}$$

$$m\left(C_v T + \frac{a_R T^4}{\rho}\right) = \varkappa \frac{d T}{d x} + \tfrac{1}{2} m u^2 - m C u + \frac{4 a_R c T^3}{3 \rho \varkappa_R} \cdot \frac{d T}{d x} - C_1 \tag{13.62}$$

where m, C and C_1 are constants of integration.

Equations (13.60) to (13.62) describe the shock wave structure (cf. Chapter IX § 3). The Rankine-Hugoniot relations across a normal shock with radiation are obtained from equations (13.60) to (13.62) by putting $d u/d x = d T/d x = 0$, i. e.

$$p_p + \frac{a_R}{3} T_0{}^4 + m u_0 = m C = p_1 + \frac{a_R}{3} T_1{}^4 + m u_1 \tag{13.63}$$

$$\begin{aligned} m\left(C_v T_0 + \frac{a_R T_0{}^4}{\rho_0}\right) - \tfrac{1}{2} m u_0{}^2 + m C u_0 &= \\ = C_1 = m\left(C_v T_1 + \frac{a_R T_1{}^4}{\rho_0}\right) &- \tfrac{1}{2} m u_1{}^2 + m C u_1 \end{aligned} \tag{13.64}$$

where subscripts 0 and 1 refer to the values in front of and behind the shock respectively. The Mach number of the shock may be defined as $M_0 = u_0/c_0$ where c_0 is the sound speed in the medium in front of the shock and c_0 is given by the following formula:

$$c_0{}^2 = \left[\frac{\partial (p_0 + \frac{1}{3} a_R T_0{}^4)}{\partial \rho_0}\right] = \frac{p_0}{\rho_0}\left[\frac{\gamma + 20(\gamma - 1)\eta + 16(\gamma - 1)\eta^2}{1 + 12(\gamma - 1)\eta}\right] \tag{13.65}$$

$$\text{where } \eta = \frac{a_R T_0^{4/3}}{p_0} = \frac{\text{radiation pressure}}{\text{gas pressure}}$$

From equations (13.56) and (13.60) to (13.62) we have the following differential equation

$$\frac{\mathrm{d}T}{\mathrm{d}u} = \frac{m\left(C_v T + a_R T^4 \frac{u}{m} - \frac{1}{2} m u^2 + m C u + C_1\right.}{\left(R_p m \frac{T}{u} - \frac{1}{3} a_R T^4\right) + m u - m C} \left(\frac{\frac{4}{3}\mu}{\varkappa + \frac{4}{3} a_R c u \frac{T^3}{m \varkappa_R}}\right) \tag{13.66}$$

Equation (13.66) gives the relation between T and u in the shock. From equation (13.66) we obtain $T(u)$ in the shock. Then from equation (13.61) or (13.62), we have $T(x)$ or $u(x)$. Numerical integration is generally used in obtaining these functions. SEN and GUESS (*13*) computed a case where the radiation pressure and radiation energy are negligible but the radiative transport of energy is not negligible in comparation with transport of thermal conduction. They found that radiation has the effect of broadening the shock front.

We may study the small wave motions for radiation-electromagnetogasdynamics in a similar manner as those discussed in Chapter X by adding the terms of radiation pressure, radiation energy and radiation flux in the fundamental equations (*9*). It was found that the radiation has influence only on the longitudinal waves by changing the ordinary sound speed into radiation sound speed c_0 (13.65). For instance, for magnetogasdynamic wave in an ideal plasma with radiation effect, equation (10.35) becomes

$$\left(c_0^2 - \frac{\omega^2}{\lambda^2}\right) \cdot \left(V_x^2 - \frac{\omega^2}{\lambda^2}\right) - V_y^2 \frac{\omega^2}{\lambda^2} = 0 \tag{13.67}$$

The Rankine-Hugoniot relations with radiation effect may also be found (*10*). The main result is that the temperature jump across the shock with radiation effect is much smaller than those without radiation effect. The influence of radiation on the velocity change across the shock is very small.

3. Particle motions of a plasma. For an exact description of the flow problem of a plasma, the collision phenomena between particles of the plasma, ions, electrons and neutral particles, should be throughly analyzed. We shall discuss briefly such collision phenomena by the kinetic theory of plasma in chapter XIV. However for a rarefied plasma such as the case at fusion temperature (*12*), the mean free path for collision of ions and electrons in a plasma will be so large that it may be larger than the dimensions of the apparatus which contains the plasma. In this case, the motion of ions and electrons of the plasma will be largely dominated by electromagnetic forces. We may apply the Newton's law of motion to an individually charged particle, ion or electron, i. e.,

$$m_r \frac{\mathrm{d}\vec{v}_r}{\mathrm{d}t} = q_r (\vec{E} + \vec{v}_r \times \vec{B}) \tag{13.68}$$

where m_r is the mass of a particle of the $r\,t\,h$ species in the plasma, $\vec{v}_r$ is its velocity vector, $q_r = Z_r e$ = electric charge on a particle of the $r\,t\,h$ species, Z_r is the number of signed elementary charge, $Z_r = -1$ for electron, e is the absolute electric charge. Equation (13.68) should be solved with the Maxwell's equations for electromagnetic fields. Many interesting results may be obtained from equation (13.68). We shall consider a few simple cases as follows:

(i) $\vec{E}$ = constant and $\vec{B} = 0$.
The particle has a constant acceleration in the direction of $\vec{E}$, i. e.,

$$\frac{d\vec{v}_r}{dt} = \frac{q_r \vec{E}}{m_r} \tag{13.68a}$$

(ii) $\vec{E} = 0$ and $\vec{B}$ = constant.
For this case, it is convenient to use cylindrical coördinates r, θ and z with z in the direction of $\vec{B}$. The final equilibrium condition will be the case that the particle has no radial velocity component, that is, axial and tangential components of velocity only. The axial velocity component V_z is determined by the initial condition of the particle and it is not affected by the magnetic field. The tangential velocity component V_θ is determined by the condition that the centrifugal acceleration is equal to the acceleration produced by the electromagnetic force, i. e.,

$$m_r \frac{V_\theta^2}{r_\lambda} = q_r v_\theta B \tag{13.69}$$

where r_λ is the radius of the circle where the particle moves in the equilibrium condition. Equation (13.69) may be written as

$$\omega_c = \frac{V_\theta}{r_\lambda} = \frac{Z_r e B}{m_r} \tag{13.70}$$

where ω_c is known as the cyclotron frequency or Larmor frequency of the charged particle m_r. If v_z is zero, the particle moves on a circle of radius r_λ; while if v_z is different from zero, the particle will move in a helical path with a radius r_λ for the helix. The radius r_λ is sometime called the Larmor radius. If we assume v_θ to be the mean molecular velocity at a given temperature T, i. e.,

$$v_\theta = \sqrt{\frac{8kT}{\pi m_r}} \tag{13.71}$$

Then the Larmor radius of this particle m_r will be

$$r_\lambda = \left(\frac{8}{\pi}\right)^{\frac{1}{2}} \frac{(k T m_r)^{\frac{1}{2}}}{Z_r e \cdot B} \tag{13.72}$$

The Larmor radius is a very important parameter in the kinetic theory of plasma.

(iii) $\vec{E}$ = constant, $\vec{B}$ = constant, and $\vec{E}$ is perpendicular to $\vec{B}$. In this case, we may define a new velocity of the particle $\vec{v}_r{}'$ such that

$$\vec{v}_r{}' = \vec{v}_r - \frac{\vec{E}\times\vec{B}}{B^2} = \vec{v}_r - \vec{v}_d \tag{13.73}$$

$$\text{where } \vec{v}_d = \frac{\vec{E}\times\vec{B}}{B^2} = \vec{i}_d \frac{E}{B} = \text{drift velocity} \tag{13.74}$$

and $\vec{i}_d$ is the unit vector perpendicular to both $\vec{E}$ and $\vec{B}$.

The velocity v_r' satisfies the following equation:

$$m_r \frac{\mathrm{d}\vec{v}_r'}{\mathrm{d}t} = q_r \vec{v}_r' \times \vec{B} \tag{13.75}$$

Hence the velocity $\vec{v}_r'$ is independent of the electric field E.

If the electric field $\vec{E}$ is not perpendicular to $\vec{B}$, we may resolve $\vec{E}$ into two components: one is perpendicular to $\vec{B}$, i. e., $\vec{E}^{\perp}$ and the other is parallel to $\vec{B}$, i. e., $\vec{E}^{\parallel}$. We use $\vec{E}^{\perp}$ in equation (13.74) to calculate the transverse drift velocity $\vec{v}_d$. The component $\vec{E}^{\parallel}$ of the electric field will produce a uniform acceleration of the particle along the lines of force.

(iv) If both $\vec{E}$ and $\vec{B}$ are not constant, the motion of the particle is very complicated. However, if the variations of the electric and the magnetic field are small, the motion of the particle will be qualitatively the same as we have discussed in the above three sections. Some of the interesting problems of plasma may be brought out by studying the particle motion in slowly varying fields such as the stability of plasma confined by magnetic fields (*13*).

In equation (13.68), we assumed that the particle is subjected to an electromagnetic force. We may generalize the motion of the particle by considering additional forces such as gravitational force. In general, the gravitational force has a similar effect as the electric field.

References

1. Burgers, J. M.: Applications of Transfer Equations to the Calculation of Diffusion, Heat Conduction, Viscosity and Electric Conductivity. I. and II. Tech. notes 124a and b. Institute for Fluid Dynamics and Applied Math. Univ. of Maryland, Mary 1958.

2. Chapman, S., and T. G. Cowling: The Mathematical Theory of Non-Uniform Gases. Cambridge University Press, 1939.

3. Engel, A. V., and M. Steenbeck: Elektrische Gasentladungen. Edwards Brothers, Inc. 1944, Julius Springer, 1932.

4. Joos, G.: Theoretical Physics. Hafner Publishing Co. New York, 1950.

5. Kemp, N. H., and H. E. Petschek: Two Dimensional Incompressible Magnetohydrodynamic Flow Across an Elliptical Solenois. Jour. Fluid Mech., vol. 4, pt. 6, Nov. 1958, pp. 553–584.

6. Lin, S. C., E. L. Resler and A. Kantrowitz: Electrical Conductivity of Highly Ionized Argon Produced by Shock Waves. Jour. App. Physics, vol. 26, No. 1, Jan. 1955, pp. 95–109.

7. Marshak, R. E.: Effect of Radiation on Shock Wave Behavior. The Physics of Fluids, vol. 1, No. 1, Jan.–Feb. 1958, pp. 24–29.

8. Pai, S. I.: The Wave Motions of Small Amplitude in a fully Ionized Plasma, Part I. Without External Magnetic Field. Rev. of Modern Phys. Oct. 1960, vol. 32, No. 4, pp. 882–887.

9. Pai, S. I., and A. I. Speth: The Wave Motions of Small Amplitude in Radiation-Electro-Magneto-Gasdynamics. Proc. 6th Midwestern Conference on Fluid Mech. Univ. of Texas, Sept. 1959, pp. 446–466.

10. Pai, S. I., and A. I. Speth: Shock Wave in Radiation-Magnetogasdynamics. Tech. Note BN-223, Institute for Fluid Dynamics and Applied Math., Univ. of Maryland, Dec. 1960.

11. Petschek, H., and S. Byron: Approach to Equilibrium Ionization behind Strong Shock Waves in Argon. Annals of Physics, vol. 1, No. 3, June 1957, pp. 270–315.

12. Post, R. F.: Controlled Fusion Research—An Application of the Physics of High Temperature Plasmas. Rev. Modern Phys., vol. 28, No. 3, July 1956, pp. 338–362.

13. Rosenbluth, M. N., and C. L. Longmire: Stability of Plasmas Confined by Magnetic Fields. Annals of Physics, vol. 1, No. 2, May 1957, pp. 120–140.

14. Saha, M. N.: Ionization in the Solar Chromosphere. Phil. Mag., vol. 40, 1920, pp. 472–488.

15. Scheibe, M.: The interaction between a Strong Shock Wave and an Inhomogeneous Magnetic Field. Ph. D. Thesis, Department of Physics, University of Maryland, 1958.

16. Sen, H. K., and A. W. Guess: Radiation Effects in Shock Wave Structure. Phys. Rev., vol. 108, No. 3, Nov. 1, 1957, pp. 560–564.

17. Spitzer, L. Jr.: Physics of Fully Ionized Gases. Interscience Publishers, N.Y., 1956.

18. Weymann, H. D.: Properties of High Temperature Argon behind Strong Shock Waves Measure with a Capacitive Probe. Tech. Note BN-62, Institute for Fluid Dynamics and Applied Math., University of Maryland, Oct. 1955.

19. Weymann, H. D.: On the Mechanism of Thermal Ionization behind Strong Shock Waves. Tech. Note BN-144, Institute for Fluid Dynamics and Applied Math., University of Maryland, July 1958.

Chapter XIV

Plasma Dynamics from Microscopic Point of View

1. Introduction. Since the plasma is composed of a large number of particles, charged and neutral, the most accurate but a more complicated method of description of the flow of a plasma is the molecular theory of plasma. However, because of many physical and mathematical difficulties, it is not possible at the present to treat the gas flow problem exactly by molecular theory. Many simplified assumptions about the molecular forces and the collision phenomena have to be made in the formulation of the theory and the resultant equations can only be solved approximately. Such a simplified molecular theory is known as the kinetic theory of gases (*5*, *8*). The generalization of the ordinary kinetic theory of gases to the case of plasma is one of the most interesting current research problems. Since it is still in the process of development, it is not possible to give a rather complete review in a short book as this. However, because of the importance of this subject to plasma dynamics, the reader should have a general idea of such a theory; hence the author will briefly discuss some of the elementary aspects of this theory so that it may serve as an introduction to this theory. For those reasers who are interested in this subject, special treatizes and papers (*2*, *3*) should be referred to.

In § 2, we shall give a brief description of the kinetic theory of gases and plasma in which the concept of the molecular velocity distribution function is introduced. The equation which governs the distribution function is known as the Boltzmann equation which will be discussed in § 3. The Maxwellian distribution which is the simplest distribution function and which is also the most well-known one will be derived in § 4.

There are definite relations between the Boltzmann equation and those of continuum mechanics which we have discussed in chapter II. In § 5, we discuss the relations of the variables in continuum mechanics and the distribution functions. With these relations, the transfer equations (§ 6) may be derived from the Boltzmann equation. The transfer equations are essentially the fundamental equations of macroscopic treatment. However from the transfer equations, we may obtain some informations for the coefficients of heat conductivity, of viscosity as well as electrical conductivity (§ 7). Then the relations between the microscopic and the macroscopic approches are established.

In § 8, we are going to solve the Boltzmann equation for plasma oscillation. Comparison with those results of chapter XIII § 5 is shown.

Finally we shall apply dimensional analysis to the Boltzmann equation in order to show various regions in the analysis of plasma dynamics as was first discussed by Kantrowitz and Petschek in § 9.

2. Molecular velocity and its distribution function. A gas or a plasma is composed of a large number of particles which are generally called molecules. The term of molecule in a general sense includes ordinary molecules, as well as atoms, ions and electrons. Thus the molecules of a gas may be all of a kind or of several kinds. The number of molecules of a gas is usually much larger than the number of kinds of molecules. For instance, at a temperature of 0° C. and one atmospheric pressure, there are about 2.69×10^{19} molecules per cubic centimeter while ordinarily there are at most a few different kinds of molecules. Since real molecules attract one another at large distance and repel one another at a small distance, the detailed interactions between molecules are very complicated. However for thin gases, the influences of molecules upon one another are rather small except when they come close to one another. Many simplifications may be assumed in the development of the kinetic theory of gases. As a result, the kinetic theory of gases has been highly developed and useful results have been obtained.

The simplest assumed picture of a gas is as follows:

The gas is assumed to be an aggregate of rapidly moving particles which are rigid bodies of zero diameter but finite mass and which are constantly colliding with one another to exchange energy. The influence of the particles on one another can be conveniently neglected until they are so close together that a collision takes place. The coarseness of the structure of the gaseous medium can be expressed by its free path which is the distance that the particles or molecules travel between collisions. Since the instantaneous velocity and density distributions of the gas are far from uniform, we can only conveniently use the statistical average of a quantity instead of the instantaneous value of a quantity. The statistical average of the free path is known as the mean free path which is one of the most important parameters in the kinetic theory of gases. If the mean free path is much smaller than the dimension of the flow field or the dimension of the body in the flow field, we have the ordinary conditions of gas dynamics in which the gas may be considered as a continuous medium. If the mean free path is not negligible when compared with the dimensions of the body in the flow field, then the effects of the discrete character of the gas must be taken into account in the study of flow problems and we have the conditions of superaerodynamics, slip flow and free molecule flow.

For a next approximation, we may assume that the molecules are still rigid bodies but surrounded by a field of forces. When two or more molecules approach closely, their paths will be influenced by the presence of the others. We call this an encounter. An encounter in this case takes the place of collision in the above simple case. For a more refined theory, we may assume that the molecules are non-rigid bodies surrounded by field of forces. Hence the concept of an encounter may still be used.

In the case of a plasma which consists of both neutral and charged particles, besides the collision or encounter within a short range just discussed, there is a kind of distant encounters which are due to the electromagnetic interaction between charged particles. This involves taking account of very small angle scattering and thus of a great number of acts of very small momentum transfer.

Furthermoore the interaction between the electromagnetic fields and the charged particles should also be taken into consideration.

If we could investigate the detailed motion of all the molecules in a gas flow, a complete description of the gas flow may be obtained. However it is not possible to do so becaus(in the first place we do not know the detailed conditions of all the gas molecules at a prescribed initial instant and in the second place since the number of the molecules of a gas is so large that it is not possible at present time to calculate the motions of all the molecules even if we know the initial conditions of all the molecules. Hence we can only conveniently consider the statistical average of the motions of the molecules of a gas.

The molecules of a gas move rapidly. The mean molecular velocity of a gas is high and of the order of sound speed in the gas. However, since collisions or encounters constantly occur, the variation of molecular velocities among molecules is expected to be large and it may vary for individual molecules from zero (at rest) up to the value of many times as large as the average. In describing such a motion, it is convenient to introduce a molecular velocity distribution function $F(\vec{r}, \vec{q}_m, t)$ where $\vec{r}$ is the vector position of a molecule in space, $\vec{q}_m$ is the molecular velocity vector at $\vec{r}$ and time t. The function F is so defined that the expectation of the number of molecules per unit volume at the position $\vec{r}$ and time t within the molecular velocity range $\vec{q}_m$ and $\vec{q}_m + d\vec{q}_m$ is

$$d\nu = F(\vec{r}, \vec{q}_m, t)\, d\vec{q}_m \tag{14.1}$$

The average number of molecule per unit volume at $\vec{r}$ and t is

$$\nu = \iiint F(\vec{r}, \vec{q}_m, t)\, d\vec{q}_m \tag{14.2}$$

If we write

$$F = \nu f \tag{14.3}$$

then

$$\iiint f(\vec{r}, \vec{q}_m, t)\, d\vec{q}_m = 1 \tag{14.4}$$

The average of any physical quantity Q of the molecules is

$$\overline{Q} = \iiint Q f\, d\xi\, d\eta\, d\zeta \tag{14.5}$$

where bar means the average value, ξ, η and ζ are the x-, y- and z-component of the molecular velocity $\vec{q}_m$ and $d\vec{q}_m$ in the integrals mean $d\xi\, d\eta\, d\zeta$.

For instance the average x-wise component of velocity of the gas flow is

$$u = \iiint f \xi\, d\xi\, d\eta\, d\zeta \tag{14.6}$$

If the gas is a mixture of several kinds of molecules, we should define a distribution function for each of the species in the mixture. We shall write the distribution function of the $s\,t\,h$ species as F_s. The relations of (14.1) to (14.6) hold true for F_s too. In other words,

$$\nu_s = \iiint F_s(\vec{r}_s, \vec{q}_{ms}, t)\, d\vec{q}_{ms} \tag{14.7}$$

where ν_s is the number density of the s-th species per unit volume in gas at $\vec{r} = \vec{r}_s$ and time t.

It should be noticed that in the most general case, the distribution function F_s must be considered as a function not only of $\vec{r}_s$ and $\vec{q}_{ms}$ but also of the internal degrees of freedom of the molecules (e. g., rotation and vibration) and of the quantum state of the electronic system. The formal analysis including these terms is straight forward. We shall not discuss it here. In this chapter we shall assume that the expression of $F_s(\vec{r}_s, \vec{q}_{ms}, t)$ is sufficient for our purpose.

3. Generalized Boltzmann equation. The main purpose of the kinetic theory of gases is to find the distribution function F_s, which is subjected to an integro-differential equation known as Boltzmann equation which has the form (*10*)

$$\frac{\partial F_s}{\partial t} + v_s^j \frac{\partial F_s}{\partial x^j} + \frac{p_s^j}{m_s} \cdot \frac{\partial F_s}{\partial v_s^j} = \frac{\delta_e F_s}{\delta t} \tag{14.8}$$

where v_s^j is the jth component of $\vec{q}_{ms}$, x^j is the jth component of $\vec{r}_s$, p_s^j is the jth component of the force acting on a molecule of sth species and m_s is the mass of a molecule of this species. The summation convention is used for the superscript j, $j = 1$, 2 or 3, but not for the subscript s. The left hand side terms denote the change of F_s from t to $t + d\,t$, while the right hand side term represents the gain or loss through collisions.

We are going to discuss briefly the collision term $\delta_e F_s/\delta t$ as follows:

We may roughly divide the collision term into two parts: the close encounter and the distant encounter.

For the close encounter term, the basic assumption in the kinetic theory of gases is that only binary collisions are considered. Under such an assumption, it is easy to show that the collision term (*4*) is

$$\left(\frac{\delta_e F_s}{\delta t}\right)_1 = 2\pi \sum_{r=1}^{N} \iint (F_s' F_r' - F_s F_r)\, g\, b\, \mathrm{d}b\, \mathrm{d}\vec{q}_{ms} = J(F_s, Fr) \tag{14.9}$$

where we assume that the gas is a mixture of N species. The summation is over all kinds of species from 1 to N. The prime refers to the F after collision, and g is the absolute initial relative velocity of the two types of molecules s and r, i. e.,

$$g = \sqrt{(v_r^1 - v_s^1)^2 + (v_r^2 - v_s^2)^2 + (v_r^3 - v_s^3)^2} \tag{14.10}$$

The parameter b is known as the impact parameter, which is the distance of closest approach of the two molecules if there would be no interaction between them. The derivation of equation (14.9) is given in standard textbook such as references (*5*) and (*8*).

For the distant encounter, the collision term is known as the Fokker-Planck mechanism. This involves taking account of very small angle scattering and thus of a great number of acts of very small momentum transfer. A certain part of the scattering is, in this way, represented as the cumulative and statistically determined result of a very large number of small-angle deflections by many other particles. This collision term due to this mechanism may be written as (*12*)

$$\left(\frac{\delta_e F_s}{\delta t}\right)_2 =$$

$$\sum_{r=1}^{N}\left[\frac{1}{2}\sum_{i,j}\frac{\partial^2}{\partial v_s^i\,\partial v_s^j}(F_s < \Delta v_s^i{}_{,r}\,\Delta v_s^j{}_{,r} >) - \sum_i \frac{\partial}{\partial v_s^i}(F_s < \Delta v_s^i{}_{,r} >)\right] = \tag{14.11}$$

$$K(F_s, F_r)$$

where, in general, for any quantity x, $\langle x_r \rangle$ is defined by

$$< x_r > = 2\pi \int_0^\infty g\, F_r \,\mathrm{d}\, q_{mr} \int_{b_c}^{b_m} x\, b\, \mathrm{d} b \tag{14.12}$$

where b_c is the lower limiting value and b_m is the upper cut-off distance for the plasma. If one takes b_m to be infinite, the integral diverges logarithmically for the case of Coulomb forces. Hence b_m must be taken to finite. Usually it may be taken as the Debye length in the plasma. The term $\langle \Delta\, v_{s,v}{}^i \rangle$ is the change of the ith velocity component due to the scattering of s and r particles. In reference (*12*), this term only has been used in Boltzmann equation to calculate the electrical conductivity of a plasma by SPITZER and HARM (*12*).

The complete collision term is the sum of equations (14.9) and (14.11), i. e.,

$$\frac{\delta_e F_s}{\delta t} = \left(\frac{\delta_e F_s}{\delta t}\right)_1 + \left(\frac{\delta_e F_s}{\delta t}\right)_2 = J(F_s, F_r) + K(F_s, F_r) \tag{14.13}$$

In general we should solve the Boltzmann equation with the collision term (14.13) simultaneously with the Maxwell equations of electromagnetic field (2.11) and (2.12) in which the electrical current density $\vec{J}$ may be written as follows:

$$\vec{J} = \sum_{s=1}^{N} e_s\, \nu_s\, \vec{q}_s = \sum_{s=1}^{N} e_s \int \vec{q}_{ms}\, F_s\, \mathrm{d}\vec{q}_{ms} \tag{14.14}$$

where e_s is the electrical charge of a particle of sth species and $\vec{q}_s$ is the average flow velocity vector of sth species. The coupling between the electromagnetic field and the matter is through the term in $\vec{J}$.

In ordinary kinetic theory of gases, we consider the Boltzmann equation with only binary collisions (14.19) and neglecting both body forces $p_s{}^j$ and electromagnetic interaction. It is well known that even the ordinary Boltzmann equation is too complicated to be solved for practical flow problems. One will not expect that the generalized Boltzmann equations for plasma with both the binary collision term and Fokker-Planck term as well as electromagnetic interaction can be solved for any practical problems in the near future. However, the generalized Boltzmann equations serve two important aspects in the study of plasma dynamics. In the first place, the fundamental equation of plasma flow from the macroscopic point of view may be derived from the Boltzmann equations as a first approximation (*3*) (cf. § 5). Thus we may obtain some information about the validity of the fundamental equations of the macroscopic description. In the second place, the Boltzmann equation may give us valuable information on the transport coefficients (*2*) such as the coefficient of electrical conductivity, etc. In the macroscopic analysis these transport coefficients are simply introduced as known functions of physical quantities of plasma dynamics. The relations between these transport coefficients and those physical quantities can not be determined in ordinary macroscopic analysis. They can only be determined experimentally or from the kinetic theory.

Even though the Boltzmann equation is too complicated to be solved for practical flow problems, it is still not an exact description for the probability distribution function. The exact description of the probability distribution function from the point of view of classical mechanics is the Liouville equation (*8*). The Boltzmann equation is only an approximation to the Liouville equation. In the Boltzmann equation, we consider the probability distribution function of a single molecule independent of all the other molecules. An improvement over the Boltzmann equation would be the consideration of distribution functions for two or more particles simultaneously. Some research in this direction has been made recently (*13*).

Another point of interest should be mentioned in connection with the collision terms in the Boltzmann equation (14.13). We have not consider the effect of radiation in the collision term. From our discussion in chapter XIII, § 2, we know that radiation may affect the number density of the various species in a plasma by a process such as photonionization. Hence radiation should affect the collision term. Little research has been done in the kinetic theory of plasma including the effect of radiation.

4. Maxwellian distribution. In geneıal it is very difficult to solve the Boltzmann equation for the distribution function F_s because the equation is non-linear. However one of the exact solution of the Boltzmann equation has been found and is known as the Maxwellian distribution function which represents a uniform and steady state of the gas. We are going to derive the Maxwellian distribution function in this section.

For simplicity, we shall make the following assumptions:

(i) We assume that the gas consists of only one kind of molecules, i. e., $N = 1$.

(ii) There is no body force on the molecules $p_s{}^j = 0$.

(iii) The collision process consists of only binary collisions, hence $K(F_s, F_r) = 0$.

(iv) The state of the gas is uniform so that $F_s = F(\vec{q}_m, t)$ which independent of $\vec{r}$.

Under these assumptions, the Boltzmann equation (14.8) becomes

$$\frac{\partial F}{\partial t} = 2\pi \int\int (F' F_1' - F F_1)\, g\, \mathrm{d}b\, \mathrm{d}b\, \vec{q}_{m1} \tag{14.15}$$

where subscript 1 refers to particles other than the one considered.

Now we define a H-function such that

$$H = \int F \cdot \log F \cdot \mathrm{d}\vec{q}_m \tag{14.16}$$

From equations (14.15) and (14.16) we have

$$\frac{\partial H}{\partial t} = 2\pi \int\int\int (1 + \log F) \cdot (F' F_1' - F F_1)\, g\, b\, \mathrm{d}b\, \mathrm{d}\vec{q}_m\, \mathrm{d}\vec{q}_{m1} \tag{14.17}$$

or

$$\frac{\partial H}{\partial t} = -\frac{\pi}{2} \int\int\int \log \left(\frac{F' F_1'}{F F_1} \right) \cdot (F' F_1' - F F_1)\, g\, b\, \mathrm{d} b\, \mathrm{d} \vec{q}_m\, \mathrm{d} \vec{q}_{m1} \quad (14.18)$$

From equation (14.18) we have

$$\frac{\partial H}{\partial t} \leqq 0 \quad (14.19)$$

because the sign of $\log (F' F'_1 / F F_1)$ and that of $(F' F'_1 - F F_1)$ are always the same. Equation (14.19) is known as the Boltzmann H-theorem.

For a steady state, $\partial H / \partial t = 0$, the integrand of equation (14.18) must vanish and we have

$$F' F'_1 = F F_1 \quad (14.20)$$

or

$$\log F' + \log F'_1 = \log F + \log F_1 \quad (14.21)$$

Equation (14.21) is known as summation invariant which can be shown to be expressible in the following form (*5*):

$$\log F = a_1 m + m \vec{a}_2 \cdot \vec{q}_m + a_3 \cdot \tfrac{1}{2} m q_m^2 \quad (14.22)$$

where a_1, $\vec{a}_2$ and a_3 are parameters independent of $\vec{r}$, $\vec{q}_m$ and t.

Equation (14.22) may be written in the following form:

$$F = a_0 \exp [- b (\vec{q}_m - \vec{v})^2] \quad (14.23)$$

where a_0 and b are constant and $\vec{v}$ is the average value of $\vec{q}_m$. In terms of the thermodynamic variables (cf. § 5), equation (14.23) becomes

$$F = \nu \left(\frac{m}{2 \pi k T} \right)^{\frac{3}{2}} \exp \left[- \frac{m (\vec{q}_m - \vec{v})^2}{2 k T} \right] = F_0 \quad (14.24)$$

$$\text{where } v = \frac{1}{\nu} \int \vec{q}_m F\, \mathrm{d} \vec{q}_m \quad (14.25)$$

Equation (14.23) or (14.24) is the well known Maxwellian distribution function which represents a uniform and steady state of the gas.

One of the method which has been extensively used in solving the Boltzmann equation is the perturbation technique in which one puts

$$F = F_0 (1 + \phi) \quad (14.26)$$

where F_0 is the Maxwellian distribution function and $|\phi| \ll 1$. Substituting equation (14.26) into the Boltzmann equation and neglecting square and higher order terms of ϕ, we may have a linearized equation for ϕ which can be solved. The detailed analysis of such a perturbation technique has been given in standard textbooks such as reference (*5*) and (*8*). We shall not discuss it here.

5. Relations between kinetic theory and conservation equations (*3, 10*). Before we derive the conservation equations from the Boltzmann equation (14.8), we have to define the relations between the distribution function F_s, the molecular velocity $\vec{q}_{ms}$ with components v_s^i and the macroscopic variables such as u_s^i, ν_s, p_s and T_s of the sth species of the plasma. We first introduce the random or peculiar velocity c_s^i and c_s^{i*} of the molecule of type s as follows:

$$v_s^i = u^i + c_s^i = u^i + w_s^i + c_s^{i*} \tag{14.27}$$

where u^i is the ith component of the mean mass flow velocity of the plasma as a whole (2.22) and w_s^i is the ith component of the diffusion velocity of the sth species in the plasma (2.23).

We use a bar to denote the mean value of a physical quantity which is defined by means of the integration over the velocity components, i. e.,

$$\overline{\Psi} = \frac{1}{\nu_s} \int F_s \Psi \, d\vec{c}_s \tag{14.28}$$

where we assume that F_s is expressed as a function of c_s^i instead of v_s^i for convenience.

The following relations of the variables used in the macroscopic treatment and the distribution function F_s are obtained:

(1) The density ρ_s is

$$\rho_s = m_s \nu_s = m_s \int F_s \, d\vec{c}_s \tag{14.29}$$

(2) The flow velocity u_s^i is

$$u_s^i = \overline{v_s^i} = \frac{1}{\nu_s} \int v_s^i F_s \, d\vec{c}_s = u^i + w_s^i \tag{14.30}$$

(3) The pressure tensor p_s^{ij} and the pressure p_s are respectively

$$p_s^{ij} = m_s \nu_s \overline{c_s^i c_s^j} \tag{14.31}$$

and

$$p_s = \tfrac{1}{3} p_s^{ii} = \tfrac{1}{3} (p_s^{11} + p_{ss}^{22} + p_s^{33}) \tag{14.32}$$

The viscous stress tensor is defined as

$$\tau_s^{ij} = - p_s^{ij} + \delta^{ij} p_s \tag{14.33}$$

where $\delta^{ij} = 0$ if $i \neq j$ and $\delta^{ij} = 1$ if $i = j$.

(4) The kinetic temperature T_s is

$$T_s = \tfrac{1}{3} \frac{m_s}{R_A \nu_s} \overline{c_s^2} \tag{14.34}$$

The temperature of the plasma as a whole is

$$T = \frac{1}{\nu} \sum \nu_s T_s \tag{14.34a}$$

(5) The energy term. The third moment of velocity is defined as

$$S_s^{ijk} = m_s \nu_s \overline{c_s^i c_s^j c_s^k} \tag{14.35}$$

The heat flux is

$$Q_{cs}^i = S_s^{ijj} = S_s^{i11} + S_s^{i22} + S_s^{i33} = q_s^{*i} + \tfrac{5}{2} p_s v_s^i \tag{14.36}$$

These partial quantities ρ_s, u_s^i etc., are those discussed in chapter II.

It should be noticed that the radiation terms such as radiation pressure and radiation energy can not be obtained from equation (14.8) because we exclude the radiation phenomena in the collision terms. Furthermore we define here the temperature, pressure and the stresses with reference to a coordinate system moving the mean flow velocity of the plasma as a whole so that all quantities for the various components are defined with regard to the same coordinate system. It is also possible to define the temperature and the stress with reference to a coordinate system moving with the mean velocity of each species, e. g.,

$$T_s = \frac{1}{3} \frac{m_s}{R_A \nu_s} \overline{c_s^{2*}} \tag{14.37}$$

We shall not discuss this approach further. Those readers who are interested in this approach may refer to reference (*3*).

From equations (14.29), (14.32) and (14.34), we have

$$p_s = \nu_s R_A T_s \tag{14.38}$$

This is the equation of state of the sth species (cf. equation (2.2)).

The force p_s^i in equation (14.8) may be written as

$$p_s^i = p_{sg}^i + p_{se}^i = m_s G^i + e_s [E^i + (\vec{q}_{ms} \times \vec{B})^i] \tag{14.39}$$

where G^i is the ith component of the gravitational acceleration.

6. Transfer equations. With the relations (14.29) to (14.36) and (14.39), we may obtain the conservation equations, sometimes known as equations of transfer (*2*), by taking moments of both side of the Boltzmann equations (14.8) with reference to the factors which are functions of the random velocity components c_s^i.

(1) Equation of continuity. For the zeroth moment, we multiply equation (14.8) by m_s and integrate the resulting equation with respect to $\vec{c}_s$. We have then

$$\frac{\partial \rho_s}{\partial t} + \frac{\partial \rho_s u_s^i}{\partial x^i} = \int m_s \left(\frac{\delta_e F_s}{\delta t}\right) d\vec{c}_s = \Delta (m_s \nu_s) \tag{14.40}$$

Equation (14.10) is identical to equation (2.3) if we put

$$\Delta (m_s \nu_s) = \sigma_s \tag{14.41}$$

In order to evaluate the collision terms, we have to assume certain model for the kinetic picture so that the collision term Δ can be evaluated. Usually such a model has to be simple so that the collision term may be easily evaluated. This is a limitation of the microscopic approach. Thus it can not be used conveniently in many complicated physical phenomena. On the other hand, the system of equations discussed in chapter II gives no information on such terms as σ_s, τ_s^{ij} etc.; but under the guidance of experimental facts one may postulate relations between the guidance of experimental facts one may postulate relations between these quantities and other physical quantities such as u_s^i, T_s etc. The modern approach to the study of plasma dynamics is to use both approaches so as to obtain a maximum amount of informations and to insure the accuracy of our postulates of the macroscopic theory as well as those of the model proposed for the kinetic theory.

For simplicity, we shall not consider the case of collisions in which the molecules s are broken up, neither such in which molecules s are formed through reactions. Hence we have

$$\Delta (m_s \nu_s) = 0 \tag{14.41a}$$

hence

$$\frac{\partial \rho_s}{\partial t} + \frac{\partial \rho_s u_s^i}{\partial x^i} = 0 \tag{14.40a}$$

(2) Equations of motion. For the first moment, we multiple equation (14.8) by $m_s c_s^i$ and integrate the resulting equation with respect to $\vec{c}_s$. We have then

$$\begin{aligned} \rho_s \frac{D u^i}{D t} + \frac{D}{D t} (\rho_s w_s^i) + \rho_s w_s^i \varepsilon + \rho_s w_s^k \frac{\partial u^i}{\partial x^k} + \frac{\partial}{\partial x^k} (p_s^{ik}) - \rho_s G^i \\ - \rho_{es} E_u^i - (\vec{J}_s \times \vec{B})^i = \int m_s c_s^i \left(\frac{\delta_e F_s}{\delta t}\right) \mathrm{d}\, c_s = \Delta_{sl}^i \end{aligned} \tag{14.42}$$

$$\text{where } \frac{D}{D t} = \frac{\partial}{\partial t} + u^k \frac{\partial}{\partial x^k}, \quad \varepsilon = \frac{\partial u^k}{\partial x^k}, \; E_u^i = E^i + (\vec{u} \times \vec{B})^i,$$

$$J_s^i = \rho_{es} u_s^i, \quad \rho_{es} = \nu_s e_s.$$

The summation of equation (14.42) with respect to s gives

$$\frac{D u^i}{D t} + \frac{\partial p^{ik}}{\partial x^k} - \rho G^i - \rho_e E_u^i - (\vec{J} \times \vec{B})^i = \sum_S \Delta_{sl}^i = 0 \tag{14.43}$$

$$\text{where } \rho = \sum_S \rho_s, \quad \rho u^i = \sum_S \rho_s u_s^i, \quad J^i = \sum_S J_s^i, \quad \rho_e = \sum_S \rho_{es}, \quad p = \sum_S p_s$$

$$\text{and} \quad p^{ik} = \sum_S p_s^{ik}.$$

Equation (14.42) with the help of equation (14.43) may be written as follows:

$$\begin{aligned} \left[\frac{D}{Dt} (\rho_s w_s^i) + \rho_s w_s^i \varepsilon + \rho_s w_s^k \frac{\partial u^i}{\partial x^k} - \frac{\partial \tau_s^{ik}}{\partial x^k} - \frac{\rho_s}{\rho} \cdot \frac{\partial \tau^{ik}}{\partial x^k}\right] \\ + \left(\frac{\partial p_s}{\partial x^i} - \frac{\rho_s}{\rho} \cdot \frac{\partial p}{\partial x^i}\right) - \left(\rho_{es} - \frac{\rho_s \rho_e}{\rho}\right) E_u^i - \left\{\left(J_s - \frac{\rho_s}{\rho} \vec{J}\right) \times \vec{B}\right\}^i = \Delta_{sl}^i \end{aligned} \tag{14.42a}$$

It should be noticed that not all the terms on the left-hand side of equation (14.42a) are of the same order, particularly when the distribution function F_s differs slightly from the Maxwellian distribution

$$F_{s0} = \frac{\nu_s}{\pi^3 a_s^3} \exp\left(-\frac{c_s^2}{a_s^2}\right) \tag{14.44}$$

where $a_s^2 = 2\, R_A\, T/m_s$, i. e.,

$$F_s = F_{s0} + F_s^{(1)} = F_{s0} (1 + \phi_s) \tag{14.45}$$

where $F_s^{(1)}$ is a small correction term. Under this condition, some of the terms in the left hand side of equation (14.42a) may be neglected. We shall discuss it in the next section.

In order to evaluate the collision term $\Delta_{sl}{}^{i}$, we shall follow BURGERS (2) by neglecting the Fokker-Planck collision terms, considering only the binary collision terms and limiting ourselves to a small correction term ϕ_s such that

$$\phi_s = A_s{}^i c_s{}^j + B_s (c_s{}^2 - a_s{}^2) + \tfrac{3}{2} B_s{}^{ij} (c_s{}^i c_s{}^j - \tfrac{1}{3} \delta^{ij} c_s{}^2) + C_s{}^i (c_s{}^2 c_s{}^i - \tfrac{5}{2} a_s{}^2 c_s{}^i) \tag{14.46}$$

where the coefficients $A_s{}^i$, B_s, etc. are functions of the coordinates r and of the time t and are as yet unknown.

Substituting F_s and ϕ_s into equations (14.30) to (14.36), we can express the coefficients $A_s{}^i$ etc., in terms of $w_s{}^i$ etc., as follows:

$$A_s{}^i = 2 \frac{w_s{}^i}{a_s{}^2} \tag{14.47a}$$

$$B_s = \frac{(T_s - T)}{(T a_s{}^2)} \tag{14.47b}$$

$$B_s{}^{ij} = -\tfrac{1}{3} \delta^{ij} B_s{}^{kk} = -2 \frac{(\tau_s{}^{ij})}{(\rho_s a_s{}^6)} \tag{14.47c}$$

$$C_s{}^i = \tfrac{8}{5} \frac{q_s{}^{*i}}{(\rho_s a_s{}^6)} \tag{14.47d}$$

The following two relations must be satisfied by $A_s{}^i$ and B_s respectively

$$\sum_S \nu_s A_s{}^i = 0 \tag{14.48a}$$

$$\sum_S \nu_s \left(\frac{B_s}{m_s}\right) = 0 \tag{14.48b}$$

It was shown in reference (2) that

$$\Delta_{sl}{}^i = -\sum_t K_{st} \left\{ w_s{}^i - w_t{}^i) - z_{st} \left(\frac{m_t}{m_0} r_s{}^i - \frac{m_s}{m_0} r_t{}^i \right) \right\} \tag{14.49}$$

where

$$K_{st} = K_{ts} = \tfrac{2}{3} \mu\, \alpha\, \nu_s \nu_t Z_{st}{}^{(11)} \tag{14.50}$$

and

$$m_0 = m_s + m_t$$

$$\eta = m_s m_t / m_0$$

$$\alpha = (2 k T / \eta)^{\frac{1}{2}}$$

$$Z_{st}{}^{(kj)} = \frac{4}{\pi^{\frac{1}{2}} \alpha^{2j+4}} \int_0^\infty \mathrm{d} g \cdot g^{2j+3} \,\mathrm{sxp} \left(-\frac{g^2}{\alpha^2} \right) S_{st}{}^{(k)} \tag{14.51}$$

$$S_{st}{}^{(k)} = S_{ts}{}^{(k)} = 2 \pi \int (1 - \cos^k \chi)\, b \,\mathrm{d} b = \text{collision cross-section} \tag{14.52}$$

$$z_{st} = 1 - \tfrac{2}{5} \cdot \frac{Z_{st}^{(12)}}{Z_{st}^{(11)}} \tag{14.53}$$

$$r_s^i = \frac{q_s^{*i}}{\nu_s k T}, \quad r_t^i = \frac{q_t^{*i}}{\nu_s k T} \tag{14.54}$$

where k is the Boltzmann constant, g is the absolute initial velocity (14.10) and χ is the angle of deflection of collision (5). Here the terms r_s^i and r_t^i give the effect of heat transfer on the equations of momentum. If we neglect these heat transfer terms, equation (14.49) becomes

$$\Delta_{sl}^i = -\sum_t K_{st} (w_s^i - w_t^i) \tag{14.55}$$

Equation (14.55) gives the first approximation for the friction coefficient as we have assumed in equation (3.19).

(3) Equation of internal energy. For the internal energy equation, we multiply equation (14.8) by $\frac{1}{2} m_s c_s^2$ and integrate the resultant equation with respect to $\vec{c}_s$. We have then

$$\begin{aligned} &\frac{D}{Dt}(\tfrac{3}{2} p_s) + \tfrac{5}{2} p_s \varepsilon - \tfrac{1}{2} (\tau_s^{ij}) \varepsilon^{ij} + \frac{\partial Q_{cs}^i}{\partial x^i} \\ &+ \rho_s \frac{D u^i}{Dt} w_s^i - \rho_s G^i w_s^i - E_u^k J_s^k = \Delta_{20} \end{aligned} \tag{14.56}$$

The summation of equation (14.56) with respect to s gives

$$\frac{D}{Dt}(\tfrac{3}{2} p) + \tfrac{5}{2} p \varepsilon - \tfrac{1}{2} \tau^{ij} \varepsilon^{ij} + \frac{\partial Q_c^i}{\partial x^i} - E_u^i J^i = 0 \tag{14.57}$$

$$\text{where } \varepsilon^{ij} = \frac{\partial u^j}{\partial x^i} + \frac{\partial u^i}{\partial x^j} - \tfrac{2}{3} \delta^{ij} \varepsilon$$

Equation (14.57) is the equation of the internal energy of the whole gas which is considered as a monatomic gas.

The collision term with the same approximation as equation (14.46) is

$$\Delta_{20} = -\sum_t 2 \nu_s \nu_t \frac{m_s m_t}{m_0^2} \alpha Z_{st}^{(11)} k (T_s - T_t) \tag{14.58}$$

(4) Equation for viscous stresses.

For the second moment in general, we multiple equation (14.8) by $m_s c_s^i c_s^j$ and integrate the resultant equation with respect to $\vec{c}_s$. We have after some manipulations:

$$\begin{aligned} &\left[\frac{D}{Dt}(p_s^{ij}) + p_s^{ij} \varepsilon + \frac{\partial}{\partial x^k}(S_s^{ijk}) + \rho_s \left\{ w_s^i \frac{D u^i}{Dt} + w_s^i \frac{D u^j}{Dt} \right\} + p_s^{ki} \frac{\partial u^j}{\partial x^k} \right] \\ &+ p_s^{kj} \frac{\partial u^i}{\partial x^k} - \rho_s (G^i w_s^i + G^j w_s^i) - (E_u^i J_s^j + E_u^j J_s^i) - (M_s^{\mathrm{II}})^{ij} = \Delta_{s2}^{ij} \end{aligned} \tag{14.59}$$

where the quantity $(M_s^{\mathrm{II}})^{ij}$ is a tensor expression depending on the viscous stresses and the magnetic induction B with the following components (2):

i, j	$(M_s^{\mathrm{II}})^{ij}$
1, 1	$2\dfrac{\rho_{es}}{\rho_s}[\tau_s^{13} B^2 - \tau_s^{12} B^3]$
2, 2	$2\dfrac{\rho_{es}}{\rho_s}[\tau_s^{12} B^3 - \tau_s^{23} B^1]$
3, 3	$2\dfrac{\rho_{es}}{\rho_s}[\tau_s^{23} B^1 - \tau_s^{13} B^2]$ (14.60)
12, or 21	$\dfrac{\rho_{es}}{\rho_s}[\tau_s^{23} B^2 - \tau_s^{13} B^1 - \tau_s^{22} B^3 + \tau_s^{11} B^3]$
13, or 31	$\dfrac{\rho_{es}}{\rho_s}[\tau_s^{12} B^1 - \tau_s^{23} B^2 - \tau_s^{11} B^2 + \tau_s^{33} B^2]$
23, or 32	$\dfrac{\rho_{es}}{\rho_s}[\tau_s^{13} B^3 - \tau_s^{12} B^2 - \tau_s^{33} B^1 + \tau_s^{22} B^1]$

Summation of equation (14.59) with respect to s gives

$$\begin{aligned}&\frac{D}{D t} p^{ij} + p^{ij}\varepsilon + \frac{\partial S^{ijk}}{\partial x^k} + p^{ki}\frac{\partial u^j}{\partial x^k} + p^{kj}\frac{\partial u^i}{\partial x^k}\\ &- (E_u^i J^j + E_u^j J^i) - \sum_S (M_s^{(\mathrm{II})ij} = \sum_S \Delta_{s2}^{ij}\end{aligned} \tag{14.61}$$

The collision term with the same approximation as equation (14.46) is (*2*):

$$\begin{aligned}\Delta_{s2} = &- \sum_t \tfrac{4}{3}\nu_s \nu_t \frac{m_s m_t}{m_0^2}\alpha Z_{st}^{(22)}\left\{\frac{\tau_t^{ij}}{\nu_t} - \frac{\tau_s^{ij}}{\nu_s} + \delta^{ij} k (T_s - T_t)\right\}\\ &+ \sum_t \tfrac{2}{5}\nu_s\nu_t\frac{m_t}{m_0}\alpha Z_{st}^{(22)}\left\{\frac{m_t}{m_0}\cdot\frac{\tau_s^{ij}}{\nu_s} + \frac{m_s}{m_0}\cdot\frac{\tau_t^{ij}}{\nu_t}\right\}\end{aligned} \tag{14.62}$$

Equation (14.59) with the help of equations (14.56) and (14.62) may be written as

$$\begin{aligned}&\Bigg[-\frac{D\tau_s^{ij}}{Dt} - \tau_s^{ij}\varepsilon + \frac{\partial}{\partial x^i}\{S_s^{kij} - \tfrac{2}{3}\delta^{ij} Q_{cs}^i\} + \rho_s\Bigg(w_s^j\frac{D u^i}{D t} + w_s^i\frac{D u^j}{D t} -\\ &- \tfrac{2}{3}\delta^{ij}\frac{D u^k}{D t} w_s^k\Bigg) - \tau_s^{ki}\frac{\partial u^j}{\partial x^k} - \tau^{kj}\frac{\partial u^i}{\partial x^k} + \tfrac{2}{3}\delta^{ij}\tau_s^{k\lambda}\frac{\partial u^\lambda}{\partial x^k}\Bigg] +\\ &+ p_s\varepsilon^{ij} - (E_u^i J_s^j + E_u^j J_s^i - \tfrac{2}{3}\delta^{ij} E_u^k J_s^k) - (M_s^{\mathrm{II}})^{ij}\\ &= -\sum_t \tfrac{4}{3}\nu_s\nu_t\frac{m_s m_t}{m_0^2}\alpha Zst^{(11)}\left\{\frac{\tau_t^{ij}}{\nu_t} - \frac{\tau_s^{ij}}{\nu_s}\right\}\\ &+ \sum_t \tfrac{2}{5}\nu_s\nu_t\frac{m_t}{m_0}\alpha Z_{st}^{(22)}\left\{\frac{m_t}{m_0}\cdot\frac{\tau_s^{ij}}{\nu_s} + \frac{m_s}{m_0}\cdot\frac{\tau_t^{ij}}{\nu_t}\right\}\end{aligned} \tag{14.59 a}$$

For engineering calculation of the flow problem, the exact equation (14.59) or (14.61) may be replaced by a simple expression with a coefficient of viscosity as shown in equation (2.34). We shall discuss such a simplification in the next section.

(5) Equation of heat flow.

If we multiply equation (14.8) by $\frac{1}{2} m_s c_s^2 c_s^i$ and integrate the resultant equation with respect to $\vec{c}_s$, we have after some manipulations:

$$\Big[\frac{D Q_{cs}^i}{D t} + Q_{cs}^i \varepsilon + \tfrac{1}{2} \cdot \frac{\partial}{\partial x^k} S_s^{kijj} + \tfrac{5}{2} \frac{p_s}{\rho_s} \cdot \frac{\partial}{\partial x^k} \tau_s^{ki} - \tau_s^{ik} \frac{D u^k}{D t} +$$
$$+ Q_{cs}^k \frac{\partial u^i}{\partial x^k} + S_s^{ikj} \frac{\partial u^j}{\partial x^k} - \tfrac{5}{2} p_s \left\{\rho_s \frac{D}{D t} (\rho_s w_s^i) + w_s^i \varepsilon + w_s^k \frac{\partial u^i}{\partial x^k}\right\}\Big] \tag{14.63}$$
$$+ \tfrac{5}{2} \frac{\nu_s k^2 T_s}{m_s} \cdot \frac{\partial T_s}{\partial x^i} + G^k \tau_s^{ik} + \frac{\rho_{es}}{\rho_s} E_u^k \tau_s^{ik} - \frac{\rho_{es}}{\rho_s} (\vec{q}^* \times \vec{B})^i = \Delta_{s3}^i$$

where $S_s^{ijk\lambda} = \nu_s m_s \overline{c_s^i c_s^j c_s^k c_s^\lambda}$.

The collission term with the same approximation as equation (14.46) is (*2*)

$$\Delta_{s3i} = \sum_t K_{st} z_{st} \frac{5 k T m_t}{2 m_0 m_s} (w_s^i - w_t^i)$$
$$- \sum_t K_{st} \frac{k T}{m_0} \left\{ z_{st}'' \frac{4 m_t}{5 m_0} (r_s^i + r_t^i) + (y_s r_s^i - y_s r_t^i) \right\} \tag{14.64}$$

$$\text{where } y_s = 3 \frac{m_s}{m_0} + \zeta_{st} \frac{m_t^2}{m_0 m_s}, \quad y_s = (3 + \zeta_{st}) \frac{m_t}{m_0},$$

$$\zeta_s = 1 + 5 z_{st} - \tfrac{7}{2} z_{st}, \quad z_{st}' = 1 - \tfrac{4}{35} \cdot \frac{Z_{st}^{(13)}}{Z_{st}^{(11)}}$$

$$z_{st}'' = \frac{Z_{st}^{(22)}}{Z_{st}^{(11)}}.$$

Similar to the viscous stresses case, the complete heat flow equation (14.63) may be replaced by a simple expression with a coefficient of heat conductivity. We shall also discuss it in the next section.

7. Expressions for the coefficients of diffusion, heat conduction and viscosity. Equations (14.42), (14.59) and (14.63) are the basic equations for the calculation of the coefficients of heat conduction, viscosity, diffusion and electrical conductivity. The detailed calculations for a general case of n-species and with gravitational force and electric and magnetic fields are very tedious. We shall not discuss them here. The readers who are interested in these calculations should refer to references (*2*, *5* and *12*). We shall discuss only a few simple cases which show how we calculate these coefficients in principle and which may serve as an introduction of these calculations.

(1) Heat conductivity in a pure gas. We consider the gas consisting of one species only and neglect the gravitational forces and electric and magnetic fields. It is well known that in ordinary gasdynamics, the total amount of heat conducted is proportional to the temperature gradient such that

$$Q_c^i = - \varkappa \frac{\partial T}{\partial x^i} \tag{14.65}$$

In the theory of continuum, no information about the coefficient of heat conductivity $\varkappa$ is given. However in the kinetic theory, we may obtain some information of $\varkappa$ from equation (14.63). Since the term containing the temperature gradient is the largest term, we may retain only this term on the right hand side of equation (14.63). For a pure gas without gravitational forcess and electric and magnetic fields, equation (14.63) reduces to

$$\tfrac{5}{2}\cdot\frac{\nu\, k^2\, T}{m}\cdot\frac{\partial T}{\partial x^i} = -\tfrac{2}{5} K\frac{k\,T}{m} z''\, r^i = -\tfrac{2}{5} K\, z''\frac{Q_c{}^i}{\nu m} \qquad (14.66)$$

where $q^{*i} = Q_c{}^i = \nu\, k\, T\, r^i$, $\eta = \dfrac{m}{2}$ and $K = \tfrac{2}{3}(k\, T\, M)^{\frac{1}{2}}\, \nu^2\, Z^{(11)}$.

Comparing equations (14.65) and (14.66) we have

$$\varkappa = \tfrac{25}{4}\cdot\frac{\nu^2\, k^2\, T}{K z''} = \frac{75\, k}{8}\left(\frac{k\,T}{m}\right)^{\frac{1}{2}}\frac{1}{Z^{(22)}} \qquad (14.67)$$

The term $Z^{(22)}$ may be evaluated by assuming some model for the kinetic picture. Then the coefficient of heat conductivity is completely known.

(2) Viscosity in a pure gas. For a pure gas it is well known that the viscous stress is proportional to the strain such that

$$\tau^{ij} = \mu\, \varepsilon^{ij} \qquad (14.68)$$

We may get some information of the coefficient of viscosity from equation (14.59). Neglecting the small terms on the right hand side of equation (14.59), we have for a pure gas

$$\tau^{ij} = \tfrac{5}{2}(k\, T\, m)^{\frac{1}{2}}\frac{\varepsilon^{ij}}{Z^{(22)}} \qquad (14.69)$$

Comparing equations (14.68) and (14.69), we have

$$\mu = \tfrac{5}{2}\cdot\frac{(k\, T\, m)^{\frac{1}{2}}}{Z^{(22)}} \qquad (14.70)$$

(3) Simplest form of the diffusion equation. In simple diffusion, without electromagnetic fields, we may define a driving force for the diffusion of the molecule s as follows:

$$\psi_s{}^i = -\left(\frac{\partial p_s}{\partial x^i} - \frac{\rho_s}{\rho}\cdot\frac{\partial p}{\partial x^i}\right) \qquad (14.71)$$

If we neglect the thermal diffusion terms $r_s{}^i$ and $r_t{}^i$ and the small terms in the square bracket in the left-hand side of equation (14.42a) we have

$$\psi_s{}^i = \sum_t K_{st}\,(w_s{}^i - w_t{}^i) \qquad (14.72)$$

The resistance coefficient K_{st} is connected with the ordinary diffusion coefficient $(D_{st})_1$ as defined in the standard textbook of CHAPMAN and COWLING (*5*) by the following formula:

$$K_{st} = \frac{\nu_s\, \nu_t\, p}{\nu^2\,(D_{st})_1} \qquad (14.73)$$

Hence we have the first approximation of the diffusion coefficient

$$(D_{st}) = \frac{3}{2}\left(\frac{k\,T}{2\,\eta}\right)^{\frac{1}{2}} \frac{1}{Z_{st}^{(11)}} \tag{14.74}$$

(4) Electric conductivity of a fully ionized gas.

Electric conductivity represents a diffusion phenomenon of charged particles in electric and magnetic fields. In general the deriving force for diffusion should be defined as follows:

$$\psi_s^{\,i} = \left(\rho_{es} - \frac{\rho_s\,\rho_e}{\rho}\right) E_u^{\,i} + \left\{\left(J_s - \frac{\rho_s}{\rho}\vec{J}\right) \times \vec{B}\right\}^i - \left(\frac{\partial\,p_s}{\gamma\,x^i} - \frac{\rho_s}{\rho}\cdot\frac{\partial\,p}{\partial\,x^i}\right) \tag{14.75}$$

We may derive a diffusion coefficient in terms of the deriving force from equation (14.42a). Since equation (14.42a) is essentially the same as the diffusion equation (3.17), the derivation of the electric conductivity from equation (14.42a) is the same as that from equation (3.17). Furthermore we see that in the derivation of electric conductivity from equation (3.17) we neglect the thermal diffusion $r_s^{\,i}$ in addition to other small terms. In reference (*11*), some general discussion of the derivation of electric conductivity from equation (14.42a) are given.

It should be pointed out that the miscroscopic approach in this chapter and the macroscopic approach of chapter II are complementary to each other. Equations such as (14.42) etc. give more detailed informations about terms such as τ_s^{ij} etc. The difficulty is the evaluation of the collision terms. Ordinarily we have to assume a simple model for the kinetic picture so that the collision terms may be evaluated. This is the limitation of the microscopic approach. Thus it can not be used conveniently in many complicated physical phenomena. On the other hand, the system of equations discussed in chapter II gives no information on such terms as σ_s, τ_s^{ij} etc.; but under the guidance of experimental facts one may postulate relations between these quantities and other physical quantities such as $u_s^{\,i}$, T_s, etc. The modern approach to the study of plasma dynamics is to use both approaches so as to obtain a maximum amount of informations and to insure the accuracy of our postulates of the macroscopic theory as well as those of the model proposed for the kinetic theory.

8. Plasma oscillations. The plasma oscillation discussed in chapter XIII § 5 may be derived from the Boltzmann equation as an initial value problem. We consider the electron oscillations in a homogeneous electrically neutral plasma in the absence of external electric and magnetic fields. We also neglect the collision terms. The Boltzmann equation for the one-dimensional motion of electrons is then

$$\frac{\partial f}{\partial t} + v\frac{\partial f}{\partial x} - \frac{e\,E}{m}\cdot\frac{\partial f}{\partial v} = 0 \tag{14.76}$$

where $f(v, x, t)$ is the distribution function of the electrons of mass m and charge-e. The average electric field E satisfies the Poisson's equation

$$\frac{\partial E}{\partial x} = e\left(\nu_0 - \int f\,\mathrm{d}\,v\right) \tag{14.77}$$

where ν_0 is the number density of the electrons in the undisturbed state.

For small disturbance, we write

$$f = f_0 + f_1 \tag{14.78}$$

where $f_0 = f_0(v)$ is the undisturbed distribution function of the electrons and $f_1 = f_1(v, x, t)$ is the perturbed distribution function which is small in comparison with f_0. The resultant linearized equations for f_1 and E are respectively

$$\frac{\partial f_1}{\partial t} + v \frac{\partial f_1}{\partial x} - \frac{e E}{m} \cdot \frac{\partial f_0}{\partial v} = 0 \tag{14.79}$$

and

$$\frac{\partial E}{\partial x} = -e \int f_1 \, \mathrm{d} v \tag{14.80}$$

We assume that at $t = 0$, an initial disturbance $f_1(v, x, 0)$ is applied to the equilibrium condition specified by $f_0(v)$. We solve equations (14.79) and (14.80) from the initial condition by the method of Fourier transform. We define the following Fourier transforms:

$$\begin{Bmatrix} f_1(v, w, k) \\ E(w, k) \end{Bmatrix} = \int_0^\infty \mathrm{d} t \int_{-\infty}^\infty \mathrm{d} x \, e^{-i(k x - w t)} \begin{Bmatrix} f_1(v, x, t) \\ E(x, t) \end{Bmatrix} \tag{14.81}$$

and the reciprocal relations

$$\begin{Bmatrix} f_1(v, x, t) \\ E(x, t) \end{Bmatrix} = \frac{1}{(2\pi)^2} \int_W \mathrm{d} w \int_{-\infty}^\infty \mathrm{d} k \, e^{i(k x - w t)} \begin{Bmatrix} f_1(v, w, k) \\ E(w, k) \end{Bmatrix} \tag{14.82}$$

where the contour W is chosen in the upper half plane parallel to the real w axis, above any poles in $f_1(v, w, k)$ or $E(w, k)$.

Multiplying equation (14.79) and integrating the resultant equation over space and time and using the relation (14.81), we have

$$f_1(v, w, k) = \frac{1}{i(k v - w)} \left[\phi(v, k) + \frac{e}{m} \cdot \frac{\partial f_0}{\partial v} E(v, k) \right] \tag{14.83}$$

where the initial value term $\phi(v, k)$ is defined as

$$\phi(v, k) = \int_{-\infty}^\infty \mathrm{d} x \, e^{-i k x} f_1(v, x, 0) \tag{14.84}$$

Similarly from equation (14.80), we have

$$E(w, k) = \frac{e}{\omega_p^2 H\left(k, \frac{w}{k}\right)} \int \frac{\phi(v, k)}{v - \frac{w}{k}} \mathrm{d} v \tag{14.85}$$

where

$$H\left(k, \frac{w}{k}\right) = \frac{k^2}{\omega_p^2} - \int \frac{G(v)}{v - \frac{w}{k}} \mathrm{d} v \tag{14.86}$$

$$\omega_p = e \sqrt{\frac{\nu_0}{m}} = \text{frequency of plasma oscillations} \tag{14.87}$$

$$G(v) = \frac{1}{\nu_0} \cdot \frac{\partial f_0}{\partial v} \tag{14.88}$$

By the relations (14.82) we may obtain $f_1(v, x, t)$ and $E(x, t)$ from equations (14.83) and (14.85) respectively. For our present analysis, it is sufficient to consider the electric field $E(x, t)$ which is

$$E(x, t) = \frac{e}{4\pi^2 \omega_p^2} \int_{-\infty}^{\infty} \mathrm{d}k \int_W \mathrm{d}w\, e^{ikx - iwt} \frac{\displaystyle\int \frac{\phi(v, k)\, \mathrm{d}v}{\left(v - \frac{w}{k}\right)}}{H\left(k, \frac{w}{k}\right)} \tag{14.89}$$

With a reasonable initial disturbance $\phi(v, k)$ the inegral will not have poles in the complex plane, the characteristic frequencies of the plasma oscillation are determined by the dispersion equation:

$$H\left(k, \frac{w}{k}\right) = 0 \tag{14.90}$$

For a Maxwellian distribution and for the case that w/k is large in comparison with v in $G(v)$, we may expand the denominator in the integral of H and obtain the following result:

$$\frac{k^2}{\omega_p^2} = -\frac{k}{w} \int G(v) \cdot \left(1 + \frac{k v}{w} + \frac{k^2 v^2}{w^2} + \dots\right) \tag{14.91}$$

From equation (14.91) we have

$$w^2 = \omega_p^2 \left(1 + \frac{3 k^2 \overline{v^2}}{\omega_p^2} + \dots\right) \tag{14.92}$$

where $\overline{v^2} = k_0 T/m = a_T^2$ = mean equare of the velocity of the distribution $f_0(v)$. Equation (14.92) with only the first two terms is essentially the same result as that given in chapter XIII § 5 except that a factor 3 replaces $\gamma = 5/3$. It is due to the fact that we assume there is no collision here; while in chapter XIII § 5, we implicitly assume that the collisions are frequent enough to insure that a local Maxwellian equilibrium is established (see reference *4*).

With the first two terms of equation (14.92) for w^2, it is easy to show that the imaginary part of w is

$$I m\, w = -\frac{\pi}{8}\, \omega_p \left(\frac{\omega_p}{k a_T}\right)^3 \exp\left(-\frac{w^2}{2 k^2 a^2_T}\right) \tag{14.93}$$

This expression shows that the plasma oscillation will damped out which is known as the Laudau damping of plasma oscillation. The plasma oscillations are damped rapidly for the wave number k of the Debye wave number $k_D = \omega_p/a_T$. For large wave lengths the damping constant vanishes exponentially so that there is no damping for such plasma oscillations.

9. Some dimensional considerations of plasma dynamics. We may apply dimensional consideration to the Boltzmann equation (14.8) in order to bring out some essential significance of various regimes in plasma dynamics. The following analysis was originally given by KANTROWITZ and PETSCHEK in reference (*9*).

We assume that the body force $p_s{}^i$ is the electromagnetic force. Hence the Boltzmann equation (14.8) may be written as follows:

$$\frac{\partial F_s}{\partial t} + v_s{}^j \frac{\partial F_s}{\partial x^j} + \frac{e_s}{m_s}[E^j + \mu_e (\vec{q}_{ms} \times \vec{H})^j] \frac{\partial F_s}{\partial v_s{}^j} = \frac{\delta_e F_s}{\delta t} \tag{14.94}$$

where e_s is the charge on a particle of sth species.

The most difficult point in the analysis of the Boltzmann equation is the evaluation of the collision term. If we are interested only in the order of magnitude of various terms, we may use the following approximation for the collision term:

$$\frac{\delta_e F_s}{\delta t} \cong \frac{F_{s0} - F_s}{\tau} \tag{14.95}$$

where F_{s0} is the Maxwellian distribution and τ is known as the relaxation time, over which a homogeneous plasma out of equilibrium relaxes toward a Maxwellian distribution in about one collision and which may be roughly written as

$$\tau \cong \frac{L_f}{U} \tag{14.96}$$

where L_f is the mean free path and U is the average speed of the particles, i. e., $U = \sqrt{8\,k\,T_s/\pi\,m_s}$.

Now we write

$$\vec{q}_{ms} = \vec{q}_s + \vec{q}_{sc} \text{ or } v_s{}^i = u_s{}^i + v_{sc}{}^i \tag{14.97}$$

Furthermore we introduce the following non-dimensional quantities:

$$\bar{t} = t\frac{U}{L}, \quad \bar{v}_s{}^i = \frac{v_s{}^i}{U}, \quad \bar{v}_{sc}{}^i = \frac{v_{sc}{}^i}{U}, \quad \bar{x}^j = \frac{x^j}{L}, \quad \vec{\bar{H}} = \frac{\vec{H}}{H} \tag{14.98}$$

where L is a typical length and H is the magnitude of $\vec{H}$.

Substituting equations (14.95) and (14.98) into equation (14.94), we have the non-dimensional form of the Boltzmann equation as follows:

$$\frac{\partial\, ln\, F_s}{\partial \bar{t}} + \bar{v}_s{}^j \frac{\partial\, ln\, F_s}{\partial \bar{x}^j} + \frac{e_s}{m_s} \cdot \frac{L}{U^2} \cdot E_u{}^j \frac{\partial\, ln\, F_s}{\partial \bar{v}_s{}^j} + \frac{L}{L_{Ls}} \mu_e (\vec{q}_{sc} \times \vec{\bar{H}})^j \frac{\partial\, ln\, F_s}{\partial \bar{v}_s{}^j} = \frac{L}{L_f}\left(\frac{F_{s0}}{F_s} - 1\right) \tag{14.99}$$

where $E_u{}^j = E^j + \mu_e (\vec{q}_s \times \vec{H})^j$ = electrical field measured in a system moving with a velocity $\vec{q}_s$.

$$L_{Ls} = \frac{m_s U}{\mu_e H e_s} = \text{Larmor radius of sts species} \tag{14.100}$$

For neutral particles, $e_s = 0$, equation (14.99) reduces to

$$\frac{\partial \ln F_s}{\partial \bar{t}} + \overline{v_s{}^j} \frac{\partial \ln F_s}{\partial \bar{x}^j} = \frac{L}{L_f} \left(\frac{F_{s0}}{F_s} - 1 \right) \tag{14.101}$$

The main parameter for equation (14.101) is the Knudsen number (L_f/L). If the Knudsen number is very small, the plasma may be considered as a continuous medium. The fundamental equations discussed in chapter II may be used to describe the flow. If $(L/L_f) \gg 1$, we have

$$\left(\frac{F_{s0}}{F_s} - 1 \right) \ll 1 \tag{14.102}$$

because the terms on the left hand side of equation (14.101) are of the order of unity. Equation (14.102) is the same as equation (14.26). It shows that for very small Knudsen number, the deviation of the distribution function F_s from the Maxwellian distribution F_{s0} is small and the perturbation technique may be used to find F_s. For first approximation, the equations of transfer obtained from the Boltzmann equation (cf. § 5) may be reduced to the same form as those discussed in chapter II and then some information about the transport coefficients may be obtained.

If the Knudsen number is very large, the collision term may be neglected. Then we have free molecule flow in which the distribution function F_s is again approximately the Maxwellian distribution.

The collision term is rather complicated when the Knudsen number is of the order of unity which is the generally the case in superaerodynamics, then the fundamental equations are much more complicated than those of Navier-Stokes equations discussed in chapter II.

For charged particles, we should use equation (14.99). In addition to the mean free path or the ordinary Knudsen number, we have a new important parameter, the Larmor radius or the Larmor number (L/L_{Ls}). Furthermore we also have a term due to the electric field, the third term of equation (14.99). For the time being, we shall assume that this term is of the same order of magnitude as the first two terms. This means that the force due to the electric field is of the same order of magnitude as the inertial force, i. e.,

$$e_s E_u = 0 \left(\frac{m_s U^2}{L} \right) \tag{14.103}$$

where E_u is a typical magnitude of $\vec{E}_u$. We shall discuss only the relative magnitudes of the mean free path L_f and the Larmor radius L_{Ls} in the plasma.

According to Kantrowitz and Petschek (*8*), we may divide plasma dynamics into the following five domains:

(i) S-domain. In this domain, we have $(L/L_f) \gg 1$ and $L/L_{Ls} \ll 1$. The continuum theory may be used and the electrical conductivity σ may be regarded as a scalar quantity (cf. chapter XIII § 6) because $L_f/L_{Ls} \ll 1$.

(ii) T-domain. In this domain, the electron Larmor radius L_{Le} becomes smaller than the mean free path and $L/L_f \gg 1$. In this region, the Hall current becomes important and the electrical conductivity should be considered as a tensor quantity. The magnetic term, i. e., the fourth term in equation (14.99) for the electron $s = e$, becomes the most important term. Since the ion Larmor radius L_{Li} is still larger than the mean free path L_f. The collision term is still dominant in the ion Boltzmann equation. The transport coefficient such as the coefficient of viscosity and thermal conductivity will be approximately the same as in the S-domain.

(iii) M-domain. In this domain, the ion Larmor radius L_{Li} becomes smaller than the mean free path. Now the magnetic term becomes the most important term in the Boltzmann equation. The Larmor radius is more important than the mean free path in determination of the transfer equation. It is expected that the equations of transfer for this case should be different from those cases where the mean free path is the dominant factor, i. e., the Navier-Stokes equations. However no simple equations of transfer have been obtained for this case yet. This is a rather important domain because it is the region where the fusion reactor will operate, and it is also a region where some astronomical phenomena occur.

(iv) EM-domain. In this domain, $L/L_{Li} \gg 1$, $L_{Li} < L_f$. In this region, the ions will move as free particles subject to the electromagnetic forces. Both the electric and the magnetic fields are important.

(v) E-domain. In this domain, $L/L_{Le} \ll 1$, $L_{Le} < L_f$. Now the main force in the Boltzmann equation is the electric force. It is a region of very low density in which many glow discharges operate.

Besides the five domains listed above, if the Debye length is larger than L, there are no interaction between gas particles. One should treat the problem by considering individual particles instead of treating it as a gas or plasma as a whole (cf. chapter XIII § 8).

In some region the radiation phenomena may become dominant factors.

From the above discussion, we see that a complete study of plasma dynamics is a very difficult problem.

References

1. Bhatnagar, P. L., E. P. Gross and M. Krook: A Model for Collision Processes in Gases I, Small Amplitude Processes in Charged and Neutral One-Component Systems. Phys. Rev., vl. 94, pp. 511–525, May 1, 1954.
2. Burgers, J. M.: The Application of Transfer Equations to the Calculation of diffusion, Heat Conduction, Viscosity and Electric Conductivity. Tech. Note BN-124a and BN-124b, Institute for Fluid Dynamics and Applied Math., University of Maryland, May 1958.
3. Burgers, J. M.: The Bridge between Particle Mechanics and Continuum Mechanics. Proc. of the Intern. Symposium on Plasma Dynamics, Woods Hole, Mass., 9–13 June, 1958, pp. 119–186. Addison-Wesley Publishing C., Inc., 1960.
4. Chandrasekhar, S.: Plasma Physics, Univ. of Chicago Press, 1960, p. 151.
5. Chapman, S., and T. G. Cowling: The Mathematical Theory of Non-Uniform Gases. Cambridge Univ. Press, 1939.
6. Grad, H.: On the kinetic theory of rarefied gases. Comm. on Pure and Applied. Math., vol. 2, 1949, pp. 331–407.
7. Gross, E. P.: Dynamics of electron beams and plasma, Symposium on Electronic wave guides. Brooklyn Polytechnic Institute, 1958.
8. Hirschefelder, J. O., C. F. Curtiss and R. B. Bird: Molecular theory of gases and liquids. John Wiley and Sons, Inc., N.Y., 1954.

9. KANTROWITZ, A. R., and H. E. PETSCHEK: An introductory discussion of magnetohydrodynamics. A symposium on magnetohydrodynamics. Ed. by R. K. M. Landshoff, Stanford Univ. Press, 1957, pp. 3–15.

10. PAI, S. I.: Plasma Dynamics from the Gasdynamic Point of View. Proc. First International Symposium of rarefied gasdynamics. Pergamon Press, pp. 394–405, 1960.

11. PIPKIN, A. C.: The electric conductivity of a partially ionized gas. Tech. Note BN-170, Institute for Fluid Dynamics and Applied Math., Univ. of Maryland, April 1959.

12. SPITZER, L. Jr., and R. HARM: Transport Phenomena in a completely ionized gas. Phys. Rev., vol. 89, pp. 977–981, 1953.

13. TCHEN, C. M.: Kinetic Equation for a Plasma with unsteady correlations. Nat. Bureau of Standards report 6274, Jan. 1, 1959.

14. TRUESDELL, C., and E. IKENBERRY: On the pressure and the flux of energy in a gas, according to Maxwell's kinetic theory I. Jour. Rat. Mech. and analysis, vol. 5, pp. 1–54 1956.

Author Index

Subject Index

Druck von Adolf Holzhausens Nfg., Wien